MIX
Papier aus verantwortungsvollen Quellen
Paper from responsible sources
FSC® C105338

Dr. Chetan Jawale

Agriculture Entomology and Pest Pesticides

Anchor Academic
Publishing

Jawale, Chetan: **Agriculture Entomology and Pest Pesticides, Hamburg, Anchor Academic Publishing 2016**

Buch-ISBN: 978-3-96067-045-2
PDF-eBook-ISBN: 978-3-96067-545-7
Druck/Herstellung: Anchor Academic Publishing, Hamburg, 2016

Bibliografische Information der Deutschen Nationalbibliothek:
Die Deutsche Nationalbibliothek verzeichnet diese Publikation in der Deutschen Nationalbibliografie; detaillierte bibliografische Daten sind im Internet über http://dnb.d-nb.de abrufbar.

Bibliographical Information of the German National Library:
The German National Library lists this publication in the German National Bibliography. Detailed bibliographic data can be found at: http://dnb.d-nb.de

All rights reserved. This publication may not be reproduced, stored in a retrieval system or transmitted, in any form or by any means, electronic, mechanical, photocopying, recording or otherwise, without the prior permission of the publishers.

Das Werk einschließlich aller seiner Teile ist urheberrechtlich geschützt. Jede Verwertung außerhalb der Grenzen des Urheberrechtsgesetzes ist ohne Zustimmung des Verlages unzulässig und strafbar. Dies gilt insbesondere für Vervielfältigungen, Übersetzungen, Mikroverfilmungen und die Einspeicherung und Bearbeitung in elektronischen Systemen.

Die Wiedergabe von Gebrauchsnamen, Handelsnamen, Warenbezeichnungen usw. in diesem Werk berechtigt auch ohne besondere Kennzeichnung nicht zu der Annahme, dass solche Namen im Sinne der Warenzeichen- und Markenschutz-Gesetzgebung als frei zu betrachten wären und daher von jedermann benutzt werden dürften.

Die Informationen in diesem Werk wurden mit Sorgfalt erarbeitet. Dennoch können Fehler nicht vollständig ausgeschlossen werden und die Diplomica Verlag GmbH, die Autoren oder Übersetzer übernehmen keine juristische Verantwortung oder irgendeine Haftung für evtl. verbliebene fehlerhafte Angaben und deren Folgen.

Alle Rechte vorbehalten

© Anchor Academic Publishing, Imprint der Diplomica Verlag GmbH
Hermannstal 119k, 22119 Hamburg
http://www.diplomica-verlag.de, Hamburg 2016
Printed in Germany

TABLE OF CONTENTS

INTRODUCTION 3

I. GENERAL CONCEPT DEFINITION AND TYPES OF PEST 5

II. SOME COMMON AGRICULTURAL PESTS WITH RESPECT TO THEIR BIOLOGICAL NAME, IDENTIFICATION MARKS, LIFE HISTORY, NATURE OF DAMAGE AND CONTROL MEASURES 13

III. NON-INSECT PESTS 38

IV. PEST CONTROL MEASURES AND PESTICIDES 53

V. PLANT PROTECTION APPLIANCES 84

REFERENCE TABLE 90

INTRODUCTION

The influence of insects on human life 'destructive and beneficial' can be traced back to prehistoric day and till now. A constant struggle is going on between men and insects in the pursuit of food. This struggle will go on forever because there does not seem to be final victory on either side. To know the destructive potentialities of some insects, one should visit a countryside invaded by locusts. Within a couple of hours of the invasion, not a blade of grass or green vegetable can be found in the area and certain scale insects can completely destroy an orchard crop. Man is himself often the victim of these insects attacks.

Agricultural Entomology
Agricultural entomology concerns the study of insects associated with various aspects of agriculture. Agricultural entomology deals with the study of both beneficial and detrimental insects. Beneficial insects include insect pollinators, such as bumble bees and honey bees; those that produce various commodities such as honey (e.g., honey bees and stingless bees; see Section 6.5.), lac (e.g., the lac insect *Kerria lacca*) and silk (e.g., the silk moth; see Section 6.6.); and natural enemies (e.g., parasitoids and predators) of agricultural pests. Insects that are detrimental to agriculture (those that cause economic losses) are commonly known as insect pests. Agricultural entomology includes ways to control insect pests, which either cause direct damage to agricultural crops or farm animals by injecting toxins and/or feeding on them; or indirect damage by serving as vectors for diseases.

The bulk of agricultural entomology deals with the control of insect pests. Half a century ago, particularly during the time of the green revolution, the philosophy of insect control was centered on eradication of the insect pest, especially through the use of persistent organic pesticides such as DDT. Chemical control was the main control method for many years; however, in 1962, a book entitled "Silent Spring", written by Rachel Carson, raised concerns about the impact of pesticides on the environment, including wildlife and human health. Since then, the philosophy of insect control has switched from eradication to maintaining insect populations below an economic injury level, that is, a point where the cost of controlling the pest equals the cost of its inflicted yield loss. Insect pest control is now conducted through integrated pest management (IPM) principles that aim to be sustainable in the use of resources and environmentally friendly. IPM requires plenty of experience and knowledge and combines all available methods of control such as biological control (e.g., use of natural enemies), cultural control (e.g., weed control, fertilization, irrigation, pruning), and mechanical or physical control (e.g., use of barriers, traps, knocking down insect pests with high water pressure), and chemical control, giving preference to products that are less harmful to humans and the environment. In IPM, control begins by knowing the insects that affect the agricultural crops. The first step in insect control is the correct identification of the insect, making taxonomy an important part of any pest control strategy. It is very important to study also the biology of the insect pest, since control may be more efficient at certain times of the insect life cycle; for example, in scale insects, chemical control is generally more effective when the insects are in their first instar (or first growth stage), just after hatching from the egg, because they are more vulnerable as they are smaller and have not developed a protective scale cover at this time. Once they grow older, scale insects produce a waxy cover that makes them less susceptible to pesticides.

In IPM, control measures should be undertaken only when necessary. This can be achieved by closely monitoring the insect populations in the field, through visual observation of crops or by the use of monitoring traps, etc. Once the relationship between damage and pest population densities are established, an economic threshold can be defined, where control measures should be started so that the potential pest population will not exceed the economic injury level. An insect should be considered a pest only when its damage exceeds the economic injury level; before that point it is a just a potential pest. There are many insects in the field, but not all insects require control as most do not cause harm and some are beneficial. Prevention is an important component of IPM programs, because by preventing potential pests there is no need of control at all. Cultural control methods such as crop rotation, the use of pest-resistant varieties, or using pest-free plantings, are some common methods of prevention. At a larger scale, prevention can be done also at points of entry, for example, quarantine inspections at airports and ports prevent the entrance of exotic pests to a country. The study of agricultural entomology involves all the basic principles of agronomy, insect ecology, life history and behavior, insect taxonomy, insect physiology, toxicology and other fields of general and applied entomology.

In India, agriculture is the main occupation of the majority of people. The major cash crops are sugarcane, cotton, citrus, groundnut, tobacco, potato etc. Apart from these coffee, tea, cashew nut, mango, grapes, oranges, various kinds of vegetables and flowers etc. Besides this major crops are sorghum, wheat, rice, maize, millets and many legumes are also cultivated on large scale as they have increasing market value but the most important natural enemies of agricultural crops are insects, plant diseases, weeds and weather conditions. Out of this insects are most successful group of animals even in the adverse climatic conditions therefore, they are greatest competitors of man in the struggle for existence. The insects which cause damage to crop plants are called as Agricultural Pest. In the present topic we will study the various kinds of pest in broad sense/view.

I. GENERAL CONCEPT DEFINITION AND TYPES OF PEST

(A) PEST – Derived from French word 'Peste' and Latin term 'Pestis' meaning plague or contagious disease.

- Pest is any animal which is noxious, destructive or troublesome to man or his interests.
- A pest is any organism which occurs in large numbers and conflict with man's welfare, convenience and profit.
- A pest is an organism which harms man or his property significantly or is likely to do so (Woods, 1976).
- Insects are pests when they are sufficiently numerous to cause economic damage (Debacli, 1964).
- Pests are organisms which impose burdens on human population by causing

(i) Injury to crop plants, forests and ornamentals.

(ii) Annoyance, injury and death to humans and domesticated animals.

(iii) Destruction or value depreciation of stored products.

- Pests include insects, nematodes, mites, snails, slugs, etc. and vertebrates like rats, birds, etc.
- Depending upon the importance, pests may be agricultural forest, household, medical, aesthetic and veterinary pests.

(B) CONCEPT OF PEST

Pest can be defined as any organism (animal or plant) whose population increases to such an extent as to cause economic losses to crops or a nuisance and health hazard to man and his livestock or possessions will be declared as a pest. The attack of pest to the agricultural crops cause economic loss to farmer. According to Edwards and Heath (1964) the pest is said to be *"Economic Pest"* if any pest causes at least 5% or more loss to the crops. The amount of damage caused to a crop is called as *"Economic damage"*. The lowest pest population density which causes damage is called as *"Economic Injury Level"*. This varies from crop to crop, area to area and season to season. For calculating the Economic Injury Level (EIL) includes four parameters:

(i) Cost of control

(ii) The market value of the crop

(iii) The yield loss attributable to a unit number of insects.

(iv) The effectiveness of the control.

Stern et. al. (1959) called the density of pest population at which control measures should be started to prevent increase in pest population from reaching the economic injury level is called as 'Economic Threshold'.

(C) TYPES OF PESTS

Insect pest are capable of feeding an almost all types of organic matter. The insects can cause damage to crop plants in the field, fruit plants, stored food and even the property. The pests can cause even health problems to man and his animals. Based on the host which they affect the pests are classified as following

1. Agricultural Crop Pests

Each and every agricultural crops are infested by number of pests that cause severe damage. Pest constitute a large number of insects attacking the various crop plants. The immature stages or adults insects are either foliage feeders or saps suckers. These insects bears chewing and sucking type of mouth parts. They may be internal feeders or borers or sub-terranean inhabitants. The important crops like jowar, bajara, wheat, cotton, sugarcane, etc. are attacked by pests like stem borer, shoot fly, Deccan wingless grasshopper, armyworms, flea beetles, aphids, leafhoppers mites, jowar midge fly, etc. The cabbage worms, semi-loopers, potato beetles, etc., possess chewing type mouth parts. They chew and shallow the external parts of the plants. While some insects i.e. blister beetle feed on pollens, petals of bajara etc. thus causing severe damage.

Sugarcane is an important cash crop cultivated widely in Maharashtra. This single crop is infested by sugarcane stem borer, shoot borer moth, root borer, pyrilla, mealy bug, scale insects etc. The cutworms, leafhoppers, potato tuber moth, epilachna beetles, mites, aphids and thrips cause injuries to potato crop and vegetables in field as well as in the storage. The cabbage leaf miner and cabbage caterpillars spoil the cabbage crop seriously. The Rhinoceros beetle, mango stem borer, brinjal fruit borer, ber fruit borer infesting the variety of fruits in the field as well as in the storage. The Thrips, Aphids, flea beetles, etc. damage grapes, resulting to great economic loss; if not controlled properly. The chickoo moth and rhinoceros beetles cause damage to chickoo and to coconuts. Thus, large number of pest organisms cause serious damage to many agricultural crop plants if not controlled properly.

Agricultural Pests can have direct or indirect effect on agricultural plants/crops as following:

(a) Direct effect

The direct effect of pests is mainly due to their feeding on the crop plants, through biting, piercing or sucking mouth parts to feed on the crop accordingly.

(i) Leaf eaters like adults and nymphs, larvae of *zonocerus,* caterpillar larvae reduce the leaf area and adversely affect the plant growth.

(ii) Stem borers and hoot flies make tunnels in the stem and disturb the conduction mechanism of the plant i.e. *Antherigona* in Maize, *Disphya* in coffee.

(iii) Some pests attack the buds at the growing points and prevent the branching and growth of the stem. i.e. *Chilo zonellus* in Jowar and *Earias* in cotton.

(iv) In some plants, the pests causes the immature fruit fall i.e. Mango fruit fly, ridge fly.

(v) The pests attack the flowers and damage the crop e.g. *Tassel* beetle of Maize.;

(vi) In some plants like maize, black maize beetle, destroy the absorbing tissue of root leading inhibit growth. Some insects consume stored food, underground tubers, etc. affect growth i.e. potato beetle.

(vii) Number of sucking insects suck cell sap from the crop plants in large quantities and results in loss of vigor of the plant. For example, Bemisia (white fly) on cotton, *Aphids* on many vegetable plants.

(viii) Some pest suck sap from flower and reduce the seed setting i.e. *coffee Lygus bugs.*

(ix) Premature nut fall is seen in coconut because of coconut hug and scales insects causes premature leaf fall.

(x) Some sucking insect inject toxins into the host body which results in distortion, gall formation, necrosis of leaves, etc. in the host e.g. Lygus bugs in cotton.

(xi) Dysdercus sucks cell sap of cotton leading vitality of plants.

(b) Indirect Effects

(i) Agricultural pests effects, delay the crop maturity and harvest. These pests not only cause loss of production but also cause *decline of* qualify, nutritional value, discoloration and market value.

(ii) The insects cause damage also act as transmission agents to transfer the pathogenic fungi, bacteria or viruses which leads secondary effects on the crop plants. For example, *platygasteri* wasp transmit coffee leaf rust, viral diseases like mosaic virus disease and curling leaf of cotton.

2. Household Pests

The diverse environments around us are very attractive to insects, including lawns, flowers, shrubs, parks, industrial complexes and dwellings. Pests that infest dwellings are commonly referred to as household pests (insects). Household insects are direct concern to man, his possessions and his immediate environment.

Insects such as cockroaches, crickets, houseflies, fruit flies, weevils, ants (red and black), and silver fish, etc. which contaminate eatable food and spoil it or transmitting disease causing agents are commonly placed under this group. The insects like cloth moths, carpet beetle, furniture beetles cause damage to property (human-possessions) are also belongs to household pest. Thus, all types of insect which are unwanted guests in the dwellings of man, which cause damage to human holdings and his health are called as household pests.

3. Storage Grain Pests

The storage of food grains has been a long practice with cultivators and traders. Considerable losses both in quality and quantity of food grains take place in storage due to number of factors. Organisms like insects, mites, rodents, fungi and bacteria are directly responsible for causing loss in stored products. It is estimated that about 74Q% stored grains are lost every year due to stored grain pest in India.

The stored food grains, seeds, fruits, nuts, etc. are infected by the internal borer insects in the Kothis, godowns and warehouses are most injurious of all insects. The borers can attack them, even during the harvesting stage in the farm land itself. The grain weevils (pulse beetle, rice weevils), moths. Red rust flour beetle, etc. cause a major damage to stored cereals (wheat, rice, bajara, barley, corn, oat, millets, etc.) and pulses (lentils, peas, beans, gram, etc.) respectively. Mainly the insects spoil the stored food grains and render them unfit for human consumption, sowing purposes. The stored *grain* pest can be differentiated into two types viz.

(i) Primary type: This group of pests cause damage to intact grains i.e. uncrushed state.

(ii) Secondary type: This group of pests feed or attack the broken or crushed grains.

4. Structural Pests

Structural pests are those harmful insects which cause damage to wooden frames, doors, furniture, fencing posts, library books, stored papers, cardboards, and all other wooden articles and components of buildings are referred as structural pests.

The termites (i.e. white ants) are colonial and social insects, feed on cellulose, mean while damage wooden material in variable form. Silver fishes food on starch material and ghee, thus damaging book bindings, wall papers, photographs and all kinds of adhesive labels. Cloth moths and carpet beetles can also be damage cloths, carpets as structural pests.

2. Medical and Veterinary Entomology

Medical entomology or human health entomology is the branch of entomology that deals with arthropods that affect human health; veterinary entomology is that branch of entomology that deals with arthropods affecting the health of nonhuman animals, particularly domesticated species. These two disciplines are often combined into a single field known as "medical and veterinary entomology". Medical and veterinary entomology involves the study of insects and other arthropods, especially arachnids, and is a broad science that includes studies on biology, ecology, morphology, taxonomy and many aspects related to disease transmission. Medical and veterinary entomology also includes pest control, parasitology, and the study of vector-borne and zoonotic diseases.

Forensic entomology is a specialist branch of medical and veterinary entomology in which insects are used as evidence in criminal investigations. Insects may be particularly important in establishing the post-mortem interval (time since death) in a homicide investigation. Insects develop at predictable rates. Since many of the insects that feed on human cadavers arrive soon after death, information about the developmental stage of these insects can be used to estimate when death actually occurred. Some types of problems caused to humans and other animals by insects and other arthropods include annoyance, envenomation, allergic reactions, invasion of host tissues, arthropod-borne diseases, food contamination, phobia for arthropods and delusory parasitosis. Envenomation occurs by the injection of venom by arthropods through bites and stings.

In Japan, the giant Asian hornet, *Vespa mandarinia*, a species native to temperate and tropical Eastern Asia, is the world's largest hornet with a body length of approximately 5 cm, a wingspan of about 7.5 cm, and a 1 cm long sting that can inject large amounts of a potent venom which has a high content of acetylcholine. Each year as many as 40 people die because of allergic reactions caused by the sting of this species.

Other problems may result when poisonous arthropods are touched or ingested; for example, the larvae of some moths can cause allergic reactions to the skin when touched, as in the case of the stinging rose caterpillar *Parasa indetermina*.

Delusory parasitosis refers to the psychosis that occurs when a person has a strong delusional belief that they are infested with parasites, when they are actually not infested with any. Each year arthropod-borne diseases and new strains of known pathogens are being discovered making the study of medical and veterinary entomology an essential field in human and animal health.

3. Arthropod-borne Diseases

Numerous human diseases are transmitted by insects and other arthropods that carry pathogens, such as bacteria, flukes, protozoa, viruses, roundworms and tapeworms, between vertebrate hosts. Insects that carry pathogens between hosts are called "vectors". Here are listed just a few of the many known vectors of human pathogens. Mosquitoes of the genus *Anopheles* are vectors of protozoans of the genus *Plasmodium*, the causal agents of malaria, which is the most deadly arthropod-borne disease, affecting about 250 million people worldwide, with about 2 million deaths accredited to this disease annually. Viruses that are transmitted through the bite of mosquitoes are commonly known as "arboviruses" (a shortening of 'arthropod-borne viruses'). The most commonly known arboviruses are those that cause dengue, yellow fever and several kinds of encephalitis (e.g., Venezuelan equine encephalitis, Western equine encephalitis, others). The yellow fever mosquito, *Aedes aegypti* is the main vector of dengue virus, yellow fever virus, and other diseases.

Fleas are capable of transmitting the bacterium, *Yersinia pestis*, the causal agent of plague. Three forms of plague (bubonic, pneumonic, and septicemic) are known to occur in humans. Plague has killed millions of people, especially in the 14th and 17th centuries and is still a problem in society, with some 5,000 cases annually.

House flies are vectors of bacteria that cause enteric diseases. For example, typhoid fever caused by *Salmonella typhi*, cholera caused by *Vibrio cholera* and shigellosis caused by *Shigella* spp. cause dysentery and diarrhea, and *Escherichia coli* causes urogenital and intestinal infections. Triatomine bugs are capable of transmitting the protozoan *Trypanosoma cruzi*, the causative agent of Chagas disease . Deer ticks may act as vectors of the bacterium, *Borrelia burgdorferi*, which causes Lyme disease in the northern hemisphere. In some parts of Africa, the tsetse flies, *Glossina* spp., are vectors of two forms of the protozoan *Trypanosoma brucei*, the causative agent of sleeping sickness or African trypanosomiasis.

4. Chagas Disease

Chagas disease, also known as American trypanosomiasis, is a potentially life-threatening illness caused by a flagellate protozoan parasite, *Trypanosoma cruzi*. The disease was named after the Brazilian physician Carlos R.J. Chagas who first described *T. cruzi* and its infection in humans in 1909. The protozoans invade the muscle cells of the digestive tract, heart, and sometimes the skeletal muscle of their hosts. The life cycle of *T. cruzi* is complex, consisting of three main developmental forms. The disease is found from the southern USA to Central America and South America and is generally transmitted to humans by the infested feces of triatomine bugs, although there are other ways of transmission (e.g., via contact of mucosal surfaces, blood transfusion, congenitally, or through organ transplant).

Triatomine bugs feeds on blood using their sucking mouthparts. As the bug feeds, it defecates in order to expel excess liquid from its gut. The feces of a bug infected with *T. cruzi* will contain the infectious parasites. The saliva of triatomine bugs contains anesthetics, anticoagulants and vasodilators that permit the insect to feed without disturbing its host. Once satiated, the insect leaves its human host and the effect of the anesthetic wears off. The bite begins to itch and the sleeping human will scratch the bug bite. The act of scratching disrupts the protective layers of the skin, allowing the protozoans found in the infected feces to penetrate the skin. Chagas disease is typical of poor rural communities, especially among people who live in houses with thatch roofs in areas where the protozoans and their triatomine bug vectors are endemic. Chagas disease affects about 18 million people in Latin America.

5. Forestry entomology

Forest entomology concerns the study of insects and other arthropods associated with forest ecosystems. It deals mostly with insect pest management, seeking to control insects that cause crown dieback or death of trees, degradation and destruction of wood, defoliation, and other problems related to the health of a forest and wood products.

Forest entomology may include studies on biodiversity, biology and ecology of insects in natural or cultivated forest ecosystems, and damage assessment to tree structures, forest stands and wood products. Insects (and other arthropods) may affect the health of a tree by interrupting its normal growth and causing stunted growth and ruining tree form.

Damage caused by insects may include perforation of tree bark, leaves and roots, which often become an entry route for pathogens (fungi, virus, others); crown dieback which consists of a substantial progressive decline in crown health, often resulting in tree death; degradation of wood quality through tunneling caused by insect feeding; staining of the wood either by feeding or by fungi carried by the tunneling insect; destruction of flowers and seeds; and decrease in photosynthetic activity and other physiological processes due to defoliation or injection of toxins, etc. On the other hand, in general, insect populations in forest ecosystems are in balance. But insects are also essential in maintaining healthy forests. Some services provided by insects in a forest include the aeration of soil through tunneling activity improving gas exchange in the root system; decomposition of organic matter of the forest floor; pollination; biological control; food source for other invertebrates and vertebrates that live in the forest; and production of food products such as honey.

II. SOME COMMON AGRICULTURAL PESTS WITH RESPECT TO THEIR BIOLOGICAL NAME, IDENTIFICATION MARKS, LIFE HISTORY, NATURE OF DAMAGE AND CONTROL MEASURES

1. RED COTTON BUG

Fig. 1: Dysdercus male and female copulating
Source: https://upload.wikimedia.org/wikipedia/commons/a/ab/Dysdercus_cingulatus,_mating.jpg

Class – Insecta
Order – Hemiptera
Family – Pyrrhocoridae
Genus – Dysdercus
Species – *cingulatus / koenigii* (Feb)

The red cotton bug has wide distribution, it is a minor pest in cotton growing region of northern India particularly Punjab and Uttar Pradesh. This pest also occurs throughout the Maharashtra state but is minor importance. It is commonly known as a "cotton stainer". (Fig. 1)

Host Plants

Cotton, bhendi, ambadi, hollyhock and several other malvaceous plants.

Identification Marks

The adult bug measures about 12-15 mm in length. The females are Longer (15 mm) than the males (12 mm). It is blood red in colour except eyes, scutellum, anal style, and antennae which are black colored. Besides, there is a black spot on each of the membranous forewings. A series of white transverse bands are present on the ventral side of the abdomen. Mouthparts are adapted for piercing and sucking. They form a straight beak or rostrum. The nymphs are smaller than adults and are wingless.

Fig. 2: Dysdercus nymph (larval form)
Source: https://upload.wikimedia.org/wikipedia/commons/0/0b/Red_cotton_bug_%28Dysdercus_koenigii%29_nymph_on_Hibiscus_lobatus_W_IMG_4064.jpg

Life Cycle

The mature female lays eggs during spring in clusters of 70-80 eggs each under the moist soil surface; fallen leaves and in crevices. The eggs are spherical, yellowish-white about 1.2 mm in length. After 7 days of incubation period and moist weather, eggs are hatched into active 1 mm long red colored nymphs which are resemble the adult except size and absence of wings (Fig. 2). The nymphs feed gregariously on the cotton bolls. The nymphs undergo 5-moults within 49-89-days to reach adult stage (Fig. 3). In winter the life of the adult is about 3-months but in summer it is varied. Pest breeds on cotton from August- November; takes shelter under leaves or debris from December-middle of March and feeds on bhendi from April-July. The life cycle of bug is completed within six to eight weeks.

Fig. 3: Dysdercus (identification marks)
source: https://upload.wikimedia.org/wikipedia/commons/c/c6/Dysdercus_Cingulatus-Fabricius_-_Red_cotton_stainer_bug_%282%29.jpg

Nature of Damage

Both nymphs and adults suck the cell sap from the leaves and tender shoots and impair the vitality of the plant. If the attack is severe, bolls open badly and the lint is of poor quality. In addition they also feed on the seeds and lower their oil content and low percentage of germination; such seeds are unfit for sowing. The lint is stained by the excreta of bugs or by their body juice as they are crushed in the ginning factories, so named cotton stainer.

Control Measures

1. Cotton field should be ploughed to expose eggs to sunlight.

2. Insects should be hand picked and killed in kerosene water.

3. The crops of bhendi should be sown as trap crop and pests collected there, should be destroyed.

4. Moistened cotton seeds should be hanged up at different places in the field where bugs congregate, they may get killed in the kerosene water.

5. Spraying of Malathion O.O5% is effective to control the pest.

6. Spraying of 1 liter endosulfan 35% EC, 0.25 liter phosphamidon = 100% EC or 1 liter Fenitrothion 100% EC per hectare is very effective or reduce pest population.

2. JOWAR STEM BORER

Class – Insecta
Order – Lepidoptera
Family– Pyralidae
Genus – *Chilo*
Species – *zonellus* = *partellus* (Swinhoe)

Jowar is the most important stable food crop of the Maharashtra state. Besides being stable food crop of the people, it also supplies very good fodder for the cattle. It is cultivated in Kharif, Rabi and also in hot weather. Jowar stem borer is one of the major pests of jowar. This pest is generally observed in the early growth of the crop and even after the ear-head formation. This pest is active throughout the year but the infestation is more noticed on rabbi and hot weather crops. The hybrid varieties are more susceptible to this pest.

Distribution

It occurs throughout India. The jowar stem borer is commonly called as spotted stalk borer or pink borer. Identification Marks The adult moth is a medium sized insect with 3 cm wing span. Its forewings are straw or light brown in colour with numerous shining brown spots on the margin and hind wings are white and papery. The caterpillars (Larvae) are dirty white in color with dark brown head with mandibulate type of mouthparts. Many dark spots are appeared on the body.

Mature caterpillars are measured about 12-20 mm in length and shows four broad and patchy strips on the body.

Fig. 4: Jowar Stem Borer

Source: https://upload.wikimedia.org/wikipedia/commons/thumb/7/79/Chilo_phragmitella.jpg/220px-Chilo_phragmitella.jpg

Fig. 5: Chilo partellus damage ('dead heart') Source: own, chetan Jawale ©

Host Plants

This is the major pest of jowar and maize but also recorded on bajra, ragi and other grasses.

Fig. 6: Host plant (Jowar / Sorghum) Source: own, chetan Jawale ©

Life Cycle

A female lays about 50-300 eggs in clusters arranged in two rows on the under surface of the leaves during April-May. Eggs are creamy white in color They hatch into the young caterpillar in about six days of incubation period. The young caterpillar feeds on tender leaves for a day or two and bores into the central shoot. The larval stage last for about 3-4 weeks and have normally five molts. Pupation takes place inside the stem and it last for about

7-10 days. The adult lives for 2-4 days. The pest is generally active from June to November and about four generations are completed in a year. The pest hibernates in the larval stage in stubbles during unfavorable period.

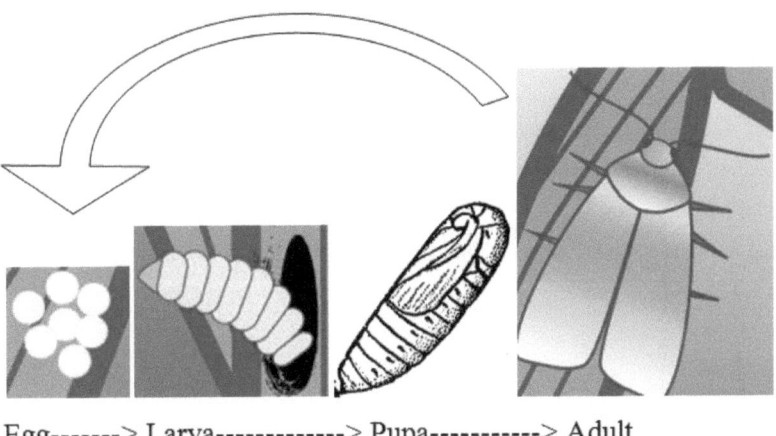

Egg-------> Larva-------------> Pupa-----------> Adult

Fig. 7: Chilo partellus Life cycle Source: own, chetan Jawale ©

Nature of Damage

Newly hatched caterpillars initially feed on the leaves causing numerous small holes in the leaf lamina and attack all parts of jowar plant except the roots. The larvae on entering the leaf, whorl and cut the leaves, which on emergence manifest characteristic pin holes, shoot holes and longitudinal streaks. At times the growing point is cut which results in drying of the central shoot and subsequently formation of dead-heart. The larvae after entering the stem, feed on the tissues (pith) and tunnels or galleries are formed.

Fig. 8: Symptoms of Damage Source: own, chetan Jawale ©

Control Measures

Cultural Method
1. Hand picking or light trapping of adult moths and collection of their eggs for destruction.
2. Burning of stubbles and trash which harbor borers and act as source of infestation for the next crop.
3. Changing the sowing and harvesting timing reduces infestation.
4. Crop rotation is another cultural practice that does not allow the pest of one crop to survive next year for lack of its host.
5. Growing resistant varieties of jowar like CHS-7, CHS-8, Indian sorghum types IS-5566, 5285 and 5613.

Chemical Method
1. For the Chilo on jowar a spray of 0.05% lindane or 0.1% endosulfan on 15 days old plants has been found effective. This may be followed after another fortnight with a second application of 1.0% lindane or 4% endosulfan granules. A third application with 0.2% carbaryl spray may be carried out, if found necessary.
2. If the crop infestation is noticed, dusting of crop in the early stage with 10% BHC at the rate of 25 kg per hectare or spraying the crop with 350-400 ml of aldrin or dieldrin in 200 liter of water helps to control the pest.

Biological Method
1. The hymenopteran, *Trichogramma minutum* is employed as egg parasite.
2. Apanteles flavipes and Bracon brevicornis as larval parasites.
3. The lady beetles, *Coccinella septempunctata* and *Menochilus.*
Sexmaculata have been recorded predating on early stages of the larvae of this pest.

3. BRINJAL FRUIT BORER

Class – Insecta

Order– Lepidoptera

Family – Pyralidae

Genus – *Leucinodes*

Species – *orbonalis* (Guenee)

Common Name: Brinjal shoot and fruit borer.

Host Plants

Brinjal (main) and other Solanaceae plants and peas (alternative). L. orbonalis is the most important and destructive pest of brinjal and has a countrywide distribution.

Identification Marks

The moths are medium sized of about 20 mm across the spread wings. The head and thorax are blackish brown. The wings are white and provided with small hairs along the apical and anal margins. A number of black, pale and light brown spots are found on the fore and hind wings of the moth. The caterpillars are pale white and about 12 mm long when fully grown.

Fig. 9: Brinjal fruit borer *Leucinodes orbonalis*

Source: https://upload.wikimedia.org/wikipedia/commons/9/9c/Leucinodes_orbonalis.jpg

Fig: 10: Brinjal fruit borer larva

Source: https://upload.wikimedia.org/wikipedia/commons/9/9c/Leucinodes_orbonalis_larva_late_instar.jpg

Life Cycle

The moth lays elongated eggs singly or in small batches, on the leaf surface, shoots and fruits. They hatch in 3-5 days. On hatching the caterpillars start boring into the shoot, leaf midrib, petiole and fruits and feeds on the internal tissues. The larva undergoes 5-moults in 10-15 days. The fifth instar larva is stout pink and measures about 1.6 cm in length. Pupation takes place in a cocoon on the plant and lasts for 6-8 days. Moth lives 2-5 days and the female lays up to 250 eggs. The larva is parasitized by Pristomerus testaceus Morl, Cremastus flauoorbitalis and Bracon species.

Nature of Damage

The larval stage is the only destructive stage. In the early stages the larvae bore into tender shoot as a result the infested shoots droop down and ultimately dry up. The larvae also bore into flower buds and developing fruits under the calyx, leaving no visible signs of infestation. The attacked fruits show holes on them plugged with excreta. In case of severe infestation in the initial stages, there may be no fruiting at all. The pinkish larvae make zigzag tunnels in the fruits and fruits are holed; such infested fruits are rendered totally unfit for human consumption. Up to 70% loss of crop is caused by this pest.

Fig. 11: Symptoms of shoot attack Source: own, chetan Jawale ©

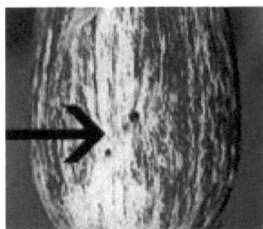

Fig. 12: Symptoms of fruit attack Source: own, chetan Jawale ©

Control Measures

1. The affected fruits and drooping shoots, containing caterpillars inside, should be clipped off and destroyed.

2. The crop should be sprayed with suspension / emulsion of any of the following insecticides.

3. The biological agencies like, Braconid wasps (*Bracois chinensis*, *Shirakia schoenobi)* and Ichneumonid wasps (*Trathela flauoorbitais*) parasitize the larvae of this pest.

4. MANGO STEM BORER

Class – Insecta
Order – Coleoptera
Family – Cerambycidae
Genus – *Batocera*
Species – *rubus* (Linn.)

The mango, the king of fruits in India, suffers from many serious pests, among them mango stem borer is the most important. It is very common in Maharashtra and Uttar Pradesh.

Host Plants
This pest is found on the planted plants like mango, fig, rubber and jack.

Identification Marks
The adult beetles are well built, large sized, measure about 5 cm long in length and brownish yellow/grey colored. It has orange yellow spots on thorax and has hard forewings (elytra); lateral spines on the prothorax and long antennae and legs. The grubs are large, yellowish white in colour, fleshy in appearance and measures about 100 x 18 mm. with black head bearing strong mandibles.

Fig. 13: Stem borer
Source: https://upload.wikimedia.org/wikipedia/commons/5/5e/Batocera_rufomaculata.jpg

Fig. 14: Grub (larva) Source: own, chetan Jawale ©

Life Cycle

The female beetle lays single egg under the loose bark or in a diseased part of trunk or in the crevices of stems. After the incubation period of 14 to 17 days the egg hatches out. The grubs on hatching penetrate into the stem or even the roots feeding on the woody tissue and make tunnels. The larval stage last for 3 to 6 months; then they pupate in the stem and remain in the pupal stage for 3 to 6 months over winter and the adults generally emerge during the monsoon. Duration of life cycle may extend 1 – 2 years.

Nature of damage

The grubs make zigzag galleries beneath the bark and tunnel into the trunks or main stems. As a result of feeding on the internal tissues, the attacked branches and stem die and wither away. Sometimes, frass and masses of refuse exude may be seen on the opening of the bored holes. In severe cases of attack, the branches may collapse and the tree may die.

Fig. 15: Damage by Mango stem borer Source: own, chetan Jawale ©

Control Measures

1. The population of grubs and pupae of stem borer can be reduced by cutting and destroying the infested branches.

2. The best way to control the grubs is to just inject borer solution (i.e. 2 parts of carbon disulphide + one part of chloroform and cresol) in the holes after which it should be closed by mud.

3. Pest population can also be effectively reduced by injecting 0.05% spray fluid of the following insecticides into the borer holes.

Insecticide	Quantity (ml.)/liter of water
DDVP (Dichlorvos) 76 EC	0.7
Endosulfan (Thiodan) 35 EC	1.5
Chiorpyriphos (Durshan) 20 EC	2.5

5. APHIDS

Class – Insecta
Order – Homoptera
Family – Aphididae

This group of insects are pests on several crops, vegetables and found in different seasons all over Maharashtra. Some species are feeding jowar and groundnut in the Kharif season, while some are on cabbage, cauliflower and safflower in the rabbi season. The cultivated crops like jowar, maize, groundnut, wal, mug, udid, bhendi, chavali, beans, lucerns, cotton and awala are infested in the Kharif season and cauliflower, cabbage, knolkhol, peas, safflower, brinjal, tondli and wheat are attacked in rabbi season. Some fruit trees like citrus also are infested by some species of aphids.

Following are the common (Aphid) pests occur all over Maharashtra on different crops:
1. Bajra and Jowar, wheat, maize - Rhophalosiphurn, maidis F.
2. Sugarcane -Aphis sacchari Z.
3. Pulses - Aphis medicaginis kie.
4. Peas - Macrosiphum pisi kalt.
5. Mug, udid, chavali, beans, wal and groundnut - Aphis cracciuora koch.
6. Lucerne - Therioaphis sexmaculata.
7. Potato and Bhendi - Myzus persicae Sulz. Aphis gossypii Glover.
8. Brinjal - Myzus persicae S.
9. Cucurbits and cotton - Aphis malvae koch. Aphis gossypii G.
10. Vegetables - Breuicoryne brassicae Linn.
11. Safflower - Dactynotus compositae Theb.
12. Citrus - Toxoptera auranti Boy. Toxoptera citricidis Kirk.

Identification Marks

Most of the aphids are generally light green or slightly yellow in colour except the safflower species which is black. They are tiny soft bodied insects; the adult is oblong about 1-2 mm long and has two projections called corniclds on the dorsal side of the abdomen. Aphids are very sluggish insects and do not move much in their wingless form and generally remain stationary after inserting their minute beaks in the leaf tissue, due to their stationary habits and small size, they are often confused by laymen as eggs of some insects. In severe attack numerous aphids are found on the undersurface of leaves and in case of jowar, they may be common in whorls also. Most aphids are wingless but in the later season or at the maturity of the crop, winged adult forms are also found. Their wings are thin, transparent, sometimes with black lines on them, and are held like a roof over the body. The Nymphs are smaller in size with greenish or brownish in colour.

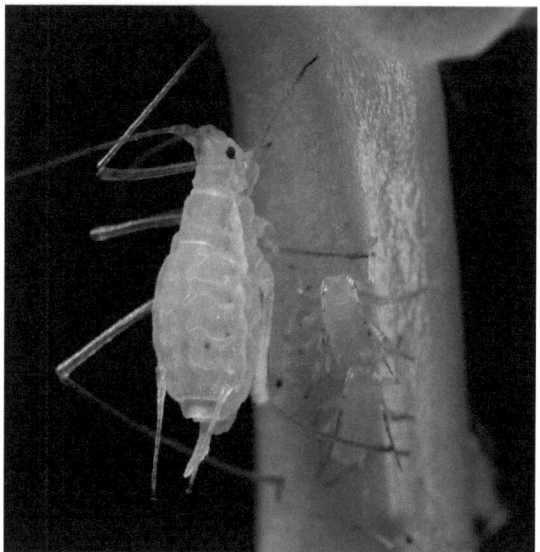

Fig. 16: Wingless form, showing adbominal cornicals
Source:https://upload.wikimedia.org/wikipedia/commons/2/20/Acyrthosiphon_pisum_%28pea_aphid%29-PLoS.jpg

Fig. 17: Winged form
Source: https://en.wikipedia.org/wiki/Myzus_persicae#/media/File:Myzus_persicae.jpg
Life history:

Both alate or winged and apterous or wingless forms produce young ones viviparously (laying young one) and parthenogenetically. All the individuals normally seen are the females, while the males are not very common or are almost absent in our climatic conditions. A single female produces a batch of 8 to 22 nymphs per day. Nymphs undergo four molts before reaching the adult stage. The life cycle is completed in 7 to 9 days and there are many generations completed in a year.

The young ones develop into mature females and thus continue the life cycle. The rate of reproduction is very rapid and the pest increases in abundance within a few weeks, resulting in the presence of hundreds of these small greenish and roundish individuals appearing on the under surface of leaves. At the end of the season as the crop approaches maturity, winged individuals appear from the wingless forms and they migrate to other crops to continue their life cycle.

Nature of Damage

Nymphs as well as adults suck the cell sap by sucking type of mouth parts in the form of a beak which is inserted in the tissue from the lower surface of the leaves and their continued feeding lead to the general yellowing of leaves and subsequent drying. As they take an excessive amount of cell sap, they also excrete a large amount of sugary solution (honeydew like substance) which comes to lounge on the leaf surface. On this sugary solution develop black sooty moulds (capnodium sp.) which blacken the surface of the leaf on which the pest feeds. This is frequently seen on the infested leaves. The sooty mould is not parasitic on the plant but remains as a saprophyte, on the surface but its presence in excessive amounts interferes with the photosynthetic activity of the plant.

These insects are known to transmit virus diseases of certain crops like cardamom and papaya.

Control Measures

1. Spraying with nicotine solution (1: 800) with five parts of soap or pyrethrum solution (1: 1000) or fish oil, rosin soap (1: 800) was recommended previously.
2. Treatment with 0.02 % endrin, methyl parathion, diazinon or 0.05 per cent Malathion gives good effect against pest.
3. 0.1 percent to 0.2 percent carbaryl or 0.02 percent phosphamidon spray is quite effective against the pest.
4. Application of insecticides like thiometon, phosphamidon, endrin + sulphur, parathion, diazinon or menazon at 0.02 percent, dimethoate at 0.03 percent concentration has also been observed to be quite effective against the pest.
5. 10% phorate or disyston granules applied at planting at 25 kg per hectare keeps crop free from aphids for 30 days.
6. DDT should not be used against aphids as it is not so effective but on the contrary it kills their parasites and predators.

Carpet beetles, belong to the family of beetles known as dermestids, are pests in warehouses, homes, museums, and other locations where suitable food exists. three species Viz. the varied carpet beetle, the furniture carpet beetle, and the black carpet beetle cause serious damage to fabrics, carpets, furs, stored food, and preserved specimens.

IDENTIFICATION AND LIFE CYCLE

All three carpet beetle species have similar life cycles (Table 1-3 below). Adults lay eggs on a larval food source such as woolen fabric or carpets or furs. Eggs hatch in about two weeks, and the larvae feed for varying periods. They prefer dark, secluded places. When ready to pupate, the larvae might burrow further into the food or wander and burrow elsewhere. They also pupate within their last larval skin if no other shelter is available. Although larvae don't make webs as clothes moths do, their shed skins and fecal pellets, which are about the size of a grain of salt, make it obvious where they have been feeding.

Carpet beetle adults don't feed on fabrics but search for pollen and nectar. They are attracted to sunlight. we often find them feeding on the flowers which produce abundant pollen. However, you can accidentally bring these pests inside on items such as cut flowers. With their rounded bodies and short antennae, carpet beetles somewhat resemble lady beetles in shape.

Tables 1-3: Life Cycle of Three Species of Carpet Beetles

Furniture Carpet Beetle

adult	larva
number of eggs laid	60
days before eggs hatch	9–16
number days for larval stage	70–94
days for pupation	14–17
weeks as an adult	4–8

Varied Carpet Beetle

adult	larva
number of eggs laid	40
days before eggs hatch	10–20
number days for larval stage	220–630
days for pupation	10–13
weeks as an adult	female 2–6; male 2–4
number of eggs laid	40

Black Carpet Beetle

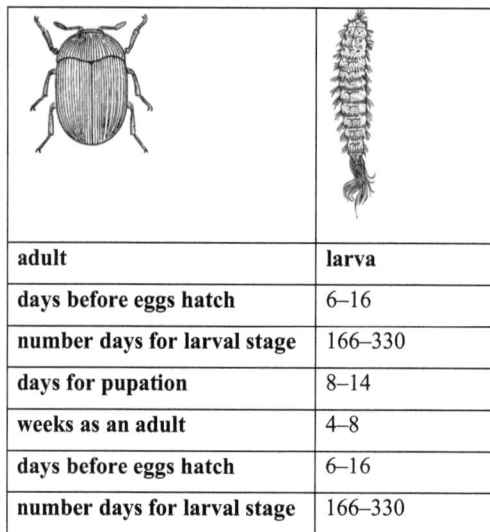

adult	larva
days before eggs hatch	6–16
number days for larval stage	166–330
days for pupation	8–14
weeks as an adult	4–8
days before eggs hatch	6–16
number days for larval stage	166–330

DAMAGE

Damage occurs during the larval stage of carpet beetles. Larvae feed in dark, undisturbed locations on a variety of dead animals and animal products such as wool, silk, leather, fur, hair brushes with natural bristles, pet hair, and feathers; occasionally they feed on stored products such as certain spices and grains. They don't feed on synthetic fibers.

one cannot identify type of beetle by looking at damage they caused, but in general carpet beetles are more likely to damage a large area on one portion of a garment or carpet while moth damage more often appears as scattered holes. Also carpet beetle larvae leave brown, shell like, bristly looking cast skins when they molt. These skins and a lack of webbing are usually good clues that carpet beetles are damaging.

MANAGEMENT

Carpet beetles are among the most difficult indoor pests to control because of their ability to find food in obscure places and to disperse widely throughout a building. Successful control depends on a combination of sanitation and exclusion. If exclusion and sanitation are successful, insecticide treatments aren't required.

Dermestids frequently fly into homes from flowers in the landscape. A few adult beetles in the house shouldn't be cause for alarm. A management program needs to be implemented only if you find larvae developing in fabrics in your home.

When carpet beetles threaten products in commercial warehouses or storage areas, a monitoring program using sticky traps baited with an appropriate pheromone (a chemical attractant an organism produces to attract others of the same species) is recommended. Sticky traps can also be used in homes where infestations are serious. Traps placed throughout a building can show where beetles are coming from; the traps are also useful for monitoring the effectiveness of control practices. Check traps once or twice a week. You can also use pheromone traps to augment other control methods if you use them to attract adult males in small confined areas. Sticky traps are also available without a pheromone; you can place these traps on windowsills to trap adults that fly to windows. Plain sticky traps are available in retail stores, while sticky traps baited with a pheromone are available from local pest control operators, pesticide supply distributors, and on the Internet. Pheromone traps are species-specific, so it is important to use one that attracts the species causing your problems.

Eliminate the Source

Eliminate accumulations of lint, hair, dead insects, and other debris that serve as food for carpet beetles. Throw out badly infested items. Remove old spider webs and bird, rodent, bee, and wasp nests, which can harbor infestations. Examine cut flowers for adult beetles before bringing the flowers inside. And be sure that window screens, doors, and vents are secure to keep carpet beetles from flying in from outdoor sources.

Regular and thorough cleaning of rugs, draperies, upholstered furniture, closets, and other locations where carpet beetles congregate is an important preventive and control technique. Frequent, thorough vacuuming is an effective way of removing food sources as well as carpet beetle eggs, larvae, and adults. After vacuuming infested areas, dispose of the bag promptly, because it can contain eggs, larvae, or adult insects.

Protect fabrics by keeping them clean; food and perspiration stains on fabrics attract carpet beetles. Thoroughly laundering washable items in hot water or dry-cleaning them will kill all stages of these insects. This is the most important method for controlling fabric pests in clothing, blankets, and other washable articles.

Regularly clean mounted animal specimens such as museum pieces or game trophies, or periodically place them in a freezer for 10 to 14 days. Inspect stored woolens, linens, and furs, and air these items annually in the sun, brushing them thoroughly. If you find an infestation, launder or dry-clean these items before returning them to storage. Be sure to seal cleaned items in a protective plastic bag or other suitable container.

Some furniture, mattresses, and pillows are stuffed with hair or feathers. When carpet beetles get into the stuffing, you can't control these insects simply by spraying the outside surface of the item. The best way to eliminate the pests is to look for a pest control, dry cleaning, or storage firm that can treat the infested item with lethal gas in a fumigation vault. Because of the potential hazards to the person applying fumigants, only licensed pest control operators can buy and use them. Proper fumigation gives quick, satisfactory control and kills all stages of fabric pests. It doesn't prevent reinfestation, however.

Protecting Items in Storage

To properly store items that are susceptible to carpet beetles, first make sure the items are pest-free and clean, then place them in an airtight container, inserting a layer of paper every few inches. On these paper layers you can place insecticide-impregnated resin strips that are labeled for control of carpet beetles on fabrics, or you can use moth balls, flakes, or crystals, which contain paradichlorobenzene (PDB), also called 1,4-dichlorobenzene. Don't let these materials come in direct contact with plastic buttons, hangers, or garment bags, since the active ingredients can cause the plastic to soften and melt into the fabric. Also be sure to keep these materials out of reach of children and pets, and don't use these materials where you store unwrapped food or allow them to come into contact with food or cooking utensils.

Resin strips, which contain dichlorvos (DDVP) as the active ingredient, are generally more effective in protecting susceptible objects in enclosed containers and provide longer control than PDB. If you use these products, be sure they are labeled for use inside homes. As these chemicals evaporate, they produce vapors that, in sufficient concentration, will slowly kill insects. The vapors build up to the required concentration only in an airtight container. If the items aren't in an airtight container, the chemicals will repel only the adults; larvae already on the items will continue to feed. Because some resin strips contain oil, don't let them come into contact with the stored item. Also don't use the resin strips in any area where people will be present for extended periods of time.

Generally, closets aren't airtight and are opened too frequently to hold in vapors. However, you can turn a seldom-used closet into a suitable storage space by sealing cracks around the door with tape or fitting the door with weather stripping. Seal cracks in walls and ceilings with putty or plastic wood.

A trunk, chest, box, or garment bag also makes a good storage container. Seal any holes or cracks, and if the lid doesn't fit tightly, seal it with tape, or wrap the entire container in heavy paper and seal it with tape.

Alternative methods for controlling carpet beetles include heating the infested object in an oven for at least 30 minutes at 120°F or higher or enclosing the object in a plastic bag and placing it in a freezer for 2 weeks at temperatures below 18°F. Before using either of these methods, consider if cold or heat will damage the object.

The effectiveness of cedar chests and closet floors made of cedar is debatable. Some cedar contains an oil that doesn't affect large larvae but can kill small ones. However, cedar loses this oil as it ages. Having a tightly constructed chest is more important in the long run than the type of wood used to make it.

Nonchemical Control

Adult carpet beetles can be captured on sticky fly paper baited with animal products and/or appropriate pheromones. Cedar products can be used to protect susceptible items. Newly hatched larvae die when exposed to cedar, but older larvae and adults are not affected. The heartwood of red cedar has a vapor that is toxic to larvae, but after cedar is more than 36 months old it is useless for control. Bags containing cedar chips should be replaced regularly to help provide control.

Plastic bags and tight containers can be used to store garments. These containers prevent adult beetles from laying eggs on or near susceptible clothing. However, if the clothing is infested, the bags will confine the infestation to just a few items.

Cold storage has been long used to protect articles attacked by carpet beetles. Clothing, coats, and sweaters stored at 40 to 42°F will be protected for long periods of time. Freezing has also been used to kill carpet beetles. Infested materials should be placed in plastic bags and loosely packed in a chest freezer at -20°F for three days. Reducing the air in the bag eliminates the formation of ice. Heat has also been used to kill or repel carpet beetle larvae. Exposure of infested items to 105°F for four hours is sufficient. Exposing infested items to hot sunlight causes larvae to abandon the fabric.

Chemical Control

Cleaning is always the best strategy; however, if you have an area or article that is infested that you can't dry clean or launder, you can spray it with an insecticide. Find a product that lists carpet beetles on its label, and closely follow the directions. Apply insecticides as spot treatments, and limit sprays to the edges of floor coverings, beneath rugs and furniture, on the floors and walls of closets, on shelving where susceptible fabrics are stored, in cracks and crevices, and in other areas that accumulate lint. Don't spray clothing or bedding.

When treating attics, wall voids, and other inaccessible places, use dust formulations such as boric acid (e.g., Eatons Answer Boric Acid Insecticidal Dust). Don't let borates come in contact with objects containing natural dyes such as some Oriental rugs, sheepskins, and bearskins. Also some dust formulations can adversely affect people who have respiratory problems; read and follow label precautions carefully. Professional fumigation might be needed when infestations are extensive, although the success rate will be lower if the fumigant can't penetrate all areas where the carpet beetles are hiding.

Closely inspect carpeted areas beneath heavy furniture and along carpet edges for infestation. If live larvae are found, spray both sides of infested carpet if at all possible, applying a lighter spray to the upper surface to reduce the possibility of staining. If the rug pad contains animal hair or wool and hasn't been treated by the manufacturer, spray it as well. It is better to wait until the rug has dried before putting any weight on it. If you are concerned that sprays might damage expensive broadloom or Oriental rugs, hire an experienced pest control operator or carpet-cleaning firm. Instead of insecticide treatment, area rugs can also be taken to dry cleaners who handle rugs.

Don't use insecticides around open flames, sparks, or electrical circuits or spray them on asphalt or tile floors. Use only a light application on parquet floors. On linoleums, first spray a small inconspicuous area and let it dry to see if staining occurs.

Applying protective sprays to furs isn't recommended. If you store furs at home during the summer, either protect them with moth crystals, flakes, or balls, or periodically shake and air them. Furs in commercial storage receive professional care, and you can insure them against damage.

Sometimes felts and hammers in pianos become infested and so badly damaged that it affects the tone and action of the instrument. Contact a piano technician, who might recommend synthetic felt replacements.

III. NON-INSECT PESTS

In addition to insect pests, many other animal pests attack the agricultural crops, stored grains, vegetables, fruits and stored products, etc., and cause damage to them. In India it has been estimated that about 5-10% of damage to agricultural crops and stored products is caused by non-insect pests. The important non-insect pests which cause damage are rats in fields and warehouses, bandicoots, wild animals like Jackals and pigs, birds, crabs, snails etc. A brief account of some non-insect pest and their control measure are described below:

1. RATS AND BANDICOOTS

Rats and Bandicoots damage the agricultural crops plants like rice, wheat, jowar, sugarcane, etc. by cutting down the plants and feeding on them. The groundnut crop is attacked by rats during harvest. Rat-burrows damage the irrigation canals also. Rats cause heavy damage to stored grains in godowns. It has been estimated that a single rat consumes about 9 kg of food grains per year. It has been found that rats spoil (due to contamination of excreta) about ten times more than it consumes. In India, it is estimated that the rat population is about 2400 million. The common species of rats recorded in India are, Rattus rattus (Black rat) and Rattus norvegicus (Brown rat). The black rats are expert climbers and prefer to live in the roofs of buildings.

Types and general characteristics of Rat (Rodent)

In several tropical countries rodents (rats and mice) cause much more loss and damage to food grain than insect pests. Three species are prevalent: the brown rat (*Rattus norvegicus*), the house mouse (*Mus musculus*) and the roof rat (*Rattus rattus*) figure 18.

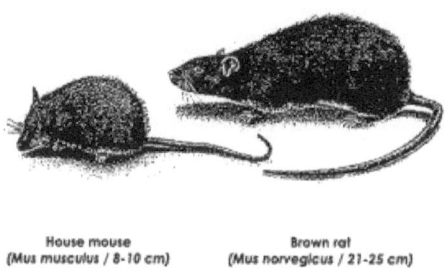

Fig. 18: Common species of rodents in storage. (Ref: AGROTEC/UNDP/OPS, 1991)

Fig 19: Bandicoots
Source: https://upload.wikimedia.org/wikipedia/commons/8/8b/Perameles_gunni.jpg

Habits and characteristics of rodents
- Usually they do same activities every day
- The follows same path over and over
- They are away of new things as for example new baits
- They can climb or jump until 60 and 90 cm
- They can swim across pools of water
- They dig and barrow holes through soil, hard structures, containers and objects

The rats as well as bandicoots are prolific breeders. Their breeding period starts at the age of 3-4 moths and continue through the year and a single female produces five litters in a year, each litter usually consists of nine to ten young ones. Thus, a pair of rat may give rise to about 800 young ones in a year. In coastal areas of Maharashtra, the coconut is reported to be damaged by rats.

They remain in the crowns of the coconut. Palm feeding on the nuts of all stages. They not only damage the grains and crop plants but also act as carries for some germs (leading to produce diseases (i.e. plague, etc.) Rattus rattus is the common species of rat, nocturnal in habit.

Bandicoots are also nocturnal in habitat they have elongated, pointed muzzle used for rooting the soil. They are pests of grains in field and store grains in houses and godowns and root crops as like rat.

Rodents loss and damage to food grain
· They damage crop in field and storage
· They eat and destroy the grain
· They foul and contaminate grain and cooked food with their feces, urine and hairs
· They can destroy buildings, structures, containers and personal clothing's and bedding
· They spillage stacked bagged foods, storage structures and food containers
· They can attack young chicks and may attempt to feed on human feet, causing sores and walking difficulties to their victims
· They may cause disease to man and animals

Rodents control methods
· By using various types of rodents traps
· By using domesticated cats
· By hunting rodents in the probably hiding places.

Control Measures
The rats and bandicoots can be killed by anyone or combined method described below:
(i) Hunting: Hunting consists of group of persons accompanied by trained cats and dogs. Men dig out the rat and bandicoot holes and when the animals come out of the holes, cats or dogs kill them. At times the burrows are flooded with the water to force the rats and bandicoots out and they can be killed mechanically.
(ii) Traps: This old method to catch the rats by various mechanical devices which are used in modified form in different locality to control the rats. This method takes much time and man power. So practically, it is very difficult to apply in larger field areas. This method is also practiced in houses to catch rats. With the help of. traps containing attractive food near their burrows, rats can be trapped. If one of the rat is caught in the trap, due to the strong instinct of self- preservation others keep away from the trap. Thus, the traps are ineffective in large scale operations.
(iii) Chemical control (Poisonings): The chemical means of rat control have proved to be most effective against Rat War. Chemicals of various nature and origin are applied with

different modes of action affecting the rat population. The following are the most commonly used chemical, preparations very effective on rats and bandicoots as poison baits.

(a) Zinc Phosphide (Zn3P2): This is very common poison used to kill the rats and bandicoots. It is a black amorphous, poisonous powder which evolves Phosphine (PH_3) gas when it comes in contact with moisture. The preparation of poison bait is as about 500 gm of moistened wheat flour is mixed with 30 gm of poison and the bait is divided into about 200 pills of equal size.

Two pills are kept near the mouth of the burrow in the evening. For bandicoots baits of little larger size are used.

(b) Barium carbonate: It is a weak poison, till it is satisfactory for use in houses. Barium carbonate is white amorphous poison used for the destruction of rats. About 500 gm of Barium carbonate is mixed with 2.5 kg of sufficiently moistened bajra flour to prepare the rat baits.

(c) White arsenic: It is one of the arsenical compound and shows same effects like zinc phosphide. It is also useful for field rats and bandicoots. The poison bait preparation involved following composition.

White arsenic – 12gm

Cooked jowar flour – 1,000 gm

Groundnut kernel – 250 gm

Water – Sufficient to make a thick paste.

Prepare small pills/cakes of equal size and keep 2 or 3 pills in each burrow and then close the burrow.

(iv) Strychnine Sulphate: As this chemical is highly toxic to human beings hence should be used only when other poisons have failed. For the preparation, dissolve 30 gm of strychnine sulphate in 60 ml of warm water. Heat 2 kg of jaggery in 500 ml of water to prepare a thick syrup. Then mix both the solutions thoroughly and add to this 14 kg gram previously soaked for 12 hours in water and make the approximately 15 grams bolls. Put 15 grams of poison bait boll in each rat burrow and then it is to be closed with mud.

(v) Kuhia seeds (strychnous nux vomica): Boil about 150 gm of the seeds in 2 liters of water so often seeds then crushed and continue boiling to get 100 gm of the extract remains. For the application follow the steps described for strychnine sulphate.

(vi) Warfarin: This is a popular rodenticide available in market. The action of this poison is slow but causing internal hemorrhage. 0.5% warfarin baits are readily available in the market. Baits must be eaten by rats for 4-5 days continuously to cause the death effect. Warfarin is highly toxic to other animals hence one should take care while handling. Rats are more suspicious to new baits so initially non- poisoning pre-baiting for 2-3 days is essential to train the rats to come to regular feeding points and on third or fourth day evening poison baits may be kept. On the next morning the dead rats must be immediately removed with uneaten baits. The composition is constitutes 1 part of warfarin ± 19 parts of food (bran) and vegetable oil. Prakash in 1976 has mentioned that poison fed rat suffers by weakness due to loss of blood and death may occur after 3 to 15 days of poisoning.

(vii) Moosh-Moosh: This is also readily available bait in the market in cake form. It is widely used to control rats from rice, sugarcane coconut and soybean fields. One or two cakes are placed in each rat hole at 5-10 meter, intervals along bunds in fields. For coconut and oil palm place two cakes on the crown (top) and at the base of each bunch of fruits. This rodenticide kills 85-100% rats with a single feeding. Application of Moosh Moosh is most economical and effective.

(viii) Ratoon, Mortem rat kill etc. are readily available in the market in cake form. These are equally effective as moosh-moosh for the control of rats and bandicoots.

(ix) Fumigation This involves the treating of rat burrows with the fumes of poisonous substances called as Cyano gas A dust or Celphos tablets. The Cyano gas A' dust is a powder, which when comes in contact with the atmospheric moisture liberates HCN gas. It is highly toxic to rats, bandicoots and other animals. Ceiphos is in tablet form containing 3 gm of Aluminium Phósphide.

Mode of Application

Apply the powder dust of Cyano gas 'A by foot pump (fumigator) into burrows having occupant rats and then close the mouth of burrow by wet mud. About 200 gm of calcium cyanide A' is enough to treat about 100 burrows, while for celphos tablets 2-3 tablets is put in each burrow, and then close it with mud. Both can be used in warehouses and godowns in the same procedure like field treatment.

The following precautions should be taken during construction of warehouses and godowns.

(a) Doors are constructed in such a way that the space left between the lower edge of the doors and floor should be less than 1 cm. The bottom of the doors must be covered with metal cuffs or plastic cuffs.

(b) Flooring of the warehouses and godowns should be without of cracks and gaps, if such are there it should be properly closed.

(c) Ventilators, windows, rain water drains, pipes, etc. should be made rat proof by covering them with wire netting (mesh with less than 1 cm pore size).

(x) Electrocuting: In many advanced countries, iron fences carrying electric charges are placed around the fields infested with rats so that rat when move across and come in contact with the electrically charged fences get killed. This method is very useful but due to inherent danger of electrocution to men and other animals hence such a device is not practicable in India.

(xi) Ultrasonic devices: Hanery Simon (1967) has recommended that ultrasonic waves are very much powerful killers of rats without any damage to human beings.

Ultrasonic equipments are now available in market which can produce sounds inaudible to human ears but such sounds which will cause pain and irritation in the ears of rats.

Rats exposed to such sounds may eventually die. Such equipments may be used both indoors and outdoors. It has been claimed by the one manufacturer of such proprietary machine that rats can be eliminated within 72 hours. Ultrasonic methods can be used only in godowns and warehouses with facilities of electricity.

Improved rodents control methods

· Prevent rodents entering by:

- solid walls structures

- use of metal sheets or chicken wire to structures walls which keep off rodents

- use of rat-guards as been before control can be achieved by baiting (poisoning). This requires training to be successful and to prevent accidents and death to humans.

· all these previous methods are strongly supported by a goods standard of hygiene and storage management.

Table 4: Recommended rodenticides

Warfarin Coumatetralyl	Anti-coagulant for mixing with dry bait or in powder form as a rodenticidal dust, or in soluble form used in drinking water for rodents
chlorophacinone	Anti-coagulant effective to rodent resistant to other anticoagulant poisons
Difenacoum Broinadiolone	
Brodifacoum	An anti-coagulant most effective to all types of rodents.

Quality and Economic Loss

Beside weight loss, which may appear small, deterioration in quality of maize due to insect attack or rats damage is also a great concern. The grain that has been damage by the insect and rats are undesirable in the market, causing great economic loss to the producer and quality loss to the consumer.

2. CRABS (PARATEIPHUSA SPP.)

The crabs have been reported to cause heavy losses to paddy crop in Ratnagiri (Konkan), Thana and Kolaba districts of Maharashtra State.

Fig 20: Rice field crab

Three species of crabs are known to damage in our country locally referred as Khekada, Chimburi and Muthya. All these crabs are polyphagous. They cut the young paddy, plants near the ground level and carry them to their burrows for feeding. They are active during night; as they are nocturnal. Besides their crop damaging activity, they prepare a series of burrow in the paddy fields due to which water is not retained in the field. Thus, the crab is major crustacean pest of paddy crop, it requires intensive control.

Control Measures

(i) Crab burrows fumigated with the cyano gas dust.

(ii) Poison baiting of 5% DDT, ± 1% dieldrin or 0.08% endrin with rice syrup or .0.5% endrin or parathion at the rate of 80-100 ml per burrow is quite effective is controlling the crabs.

3. SNAILS AND SLUGS

These are non-insect invertebrate pests and are herbivorous in habit. The land snails and slugs damage gardens, orchards, green houses and mushroom beds as they feed on succulent parts of seedlings and mature plants. Helix spp feed on living vegetable matter like leaves and fruits during night, Pila suppose to damage paddy fields, and African snail Achatina fulica is serious pest of fruits, vegetables and ornamental plants in coastal areas of Orissa, W.B., Assam, Tamil Nadu and Kerala.

Fig. 21: Lymnaea
Source: https://upload.wikimedia.org/wikipedia/commons/9/9c/Lymnaea_auricularia2pl.jpg

Snails can cause damage to crops, particularly at the early stages and thus be agricultural pests. Besides this it may be vectors of diseases such as Schistosomiasis. from veterinary and medical point of view.

Control Measures

The best known chemical control of snails is the use of poison bait with metaldehyde, a polymer of acetaldehyde. It is toxic to snails by contact as well as ingestion. The chemical immobilizes snails and copious slime is exuded out of the body of snails and they die of dehydration. It is applied as a bait mixed with bran and the dosage is very low. About 400 gm of metaldehyde mixed with 30 kg of bran is sufficient with 30 kg of bran is sufficient per hectare will control slugs. Metaldehyrde have low mammalian toxicity.

Fig. 22: Helix

Source: https://upload.wikimedia.org/wikipedia/commons/6/6d/Common_snail.jpg

Fig. 23: African giant snail

Source: https://upload.wikimedia.org/wikipedia/commons/f/fc/Achatina_fulica_Thailand.jpg

Fig. 24: Banana Slug

Source: https://upload.wikimedia.org/wikipedia/commons/a/ac/Banana_slug_-_Rialto_Beach,_Olympic_National_Park,_Washington_-_25_July_2013.jpg

DNOC or dinitro-O-cyclo-hexyphenol reported to be very effective against snails when used as herbicide. Copper sulphate and N-trityl morpnoline (frescon) have been found very useful against snails when they are spread on meadows harboring these animals.

4. BIRDS

Birds are often considered to be beneficial for the control of insects as they often live on insects and reduce the population. But birds may cause damage to early season and off season crops. They have been found to be very destructive to the cobs of maize or the heads of sunflower or ripening fruits. They may cause damage to tender seedlings of vegetables.

Fig. 25: House sparrow, Passer domesticus. (Male).

Source:

https://upload.wikimedia.org/wikipedia/commons/f/fe/House_Sparrow_%28Passer_domesticus_indicus%29.jpg

Many species of birds are found throughout India, out of which some birds are considered harmful to agricultural crops.

Some of them are as following:

(i) Crow (Corvus Splendens Vieillot): Crows cause considerable damage to ripe fruits in orchards and also ripening grains of maize and fruits. The crows are particularly attracted to the grains when they are exposed on a cob. They may prove a menace to the successful growth of field crops as well as harvest of fruits. They are often seen in flocks in maize and other fields.

(ii) Sparrows (Passer domesticus): The flocks of sparrows is a great menace to various field crops like Jowar, bajara, wheat, maize, etc. mainly in the seed setting stage. They also threaten mulberry and many other small sized juicy fruits and fruit buds they visit the ripening fruit fields, particularly. Those of wheat in the spring season, and cause much damage both by feeding and causing the grains to shed.

Damage by House sparrows

House sparrows consume grains in fields and in storage. They do not move great distances into grain fields, preferring to stay close to the shelter of hedgerows. Localized damage can be considerable since sparrows often feed in large numbers over a small area.

Sparrows damage crops by pecking seeds, seedlings, buds, flowers, vegetables, and maturing fruits. They interfere with the production of livestock, particularly poultry, by consuming and contaminating feed. Because they live in such close association with humans, they are a factor in the dissemination of diseases (chlamydiosis, coccidiosis, erysipeloid, Newcastle's, parathypoid, pullorum, salmonellosis, transmissible gastroenteritis, tuberculosis, various encephalitis viruses, vibriosis, and yersinosis), internal parasites (acariasis, schistosomiasis, taeniasis, toxoplasmosis, and trichomoniasis), and household pests (bed bugs, carpet beetles, clothes moths, fleas, lice, mites, and ticks).

In grain storage facilities, fecal contamination probably results in as much monetary loss as does the actual consumption of grain. House sparrow droppings and feathers create janitorial problems as well as hazardous, unsanitary, and odoriferous situations inside and outside of buildings and sidewalks under roosting areas. Damage can also be caused by the pecking of rigid foam insulation inside buildings. The bulky, flammable nests of house sparrows are a potential fire hazard. The chattering of the flock on a roost is an annoyance to nearby human residents.

Nestlings are primarily fed insects, some of which are beneficial and some harmful to humans. Adult house sparrows compete with native, insectivorous birds. Martins and bluebirds, in particular, have been crowded out by sparrows that drive them away and destroy their eggs and young. House sparrows generally compete with native species for favored nest sites.

(iii) Parrots (Psittacula spp.): About eight species of parrots have been recorded in India. Out of these species, large Indian parakeet (P. eupatria) is very common in Maharashtra. This species causes heavy damage to orchards by eating fruits and also spoiling the fruits by cutting it with beak. The parakeets are among the most wasteful a destructive birds. They gnaw at and cut into bits all sorts of near-ripe fruits such as guava, ber, mango, plums, peaches, etc.

Control of Birds

Various methods are employed which include covering by nets using scaring devices, reducing their population by shooting, trapping and use of chemicals.

(a) Trapping the birds in nets or catching them with the help o sticky substance 'Lassa'.

(b) A piece of Chapatti dipped in 0.04% parathion and placed or top of roof is a good bait for crows.

(c) Parrots and sparrows are repelled by spraying 0.6% thiurun' on wheat crops at milk stage.

(d) Scaring devices using mechanical, acoustic and visual means are normally employed, i.e. Beating of drums to produce sounds is still in vogue in many parts of the country particularly in the harvest season.

(e) Fire crackers placed at regular intervals along a cotton rope. The rope burns from one end and ignites the crackers at regular interval which produce sounds and scare away the birds.

Fig. 26: *Psittcula eupatria*

Source:https://upload.wikimedia.org/wikipedia/commons/8/8c/Alexandrine_Parakeet_%28Psittacula_eupatria%29,_Jurong_Bird_Park,_Singapore_-_20090613.jpg

Fig. 27: *Corvus splendens*
Source: https://c2.staticflickr.com/4/3449/3374307680_f448c2daec_b.jpg

(f) Loud sounds due to the burning of acetylene gas produced at intervals are utilized to scare away birds and small animals.

(g) Birds may be scared by display of scare crows, dead birds, visually attractive flags, etc.

For the preservation of bird species, trapping, shooting and use of poisonous chemicals as baits has not been favored in many places, instead scaring is adopted.

IV. PEST CONTROL MEASURES AND PESTICIDES

1. Chemical control:

Chemicals used for controlling pests can be divided into different groups based on the nature of pests to be controlled like:

(i) Acaricides: Chemicals used to control of mites, ticks, etc.

(ii) Algicides: Used for the destruction of algae and other aquatic vegetation.

(iii) Antispetics: For the protection of non-metals from microorganisms.

(iv) Arboricides: Destruction of undesirable arboreal and bushy vegetation.

(v) Bactericides: For the control of bacterial diseases.

(vi) Fungicides: For the control of plant diseases caused by various fungi.

(vii) Herbicides: For the control of weeds.

(viii) Insecticides: For the control of harmful insects/pests.

(ix) Limacides or Molluscides: For the control of molluscs.

(x) Nematocides: For the control of nematodes.

(xi) Rodenticides: For the control of rodents and other vertebrate animals.

Insecticides:

Insecticides can be divided into the following groups according to the nature of their penetration or mode of action into the insect body as following:

(1) Stomach poisons. (2) Contact poisons. (3) Systematic poisons. (4) Fumigants.

(1) Stomach poisons:

This type of poison act through the stomach by taken in along the food and absorbed through the digestive tract. These are commonly mixed with dust or its solution is spread up and down in and above the plant, the insect die of eating them. Generally, stomach poisons are applied against insects possessing chewing type of mouth parts. But they are also used for insects with sponging, siphoning, lapping or sucking mouth parts under certain condition.

Good stomach poison should possess the following properties.

(i) It should be stable, cheap and available in large quantities.

(ii) It should not be distasteful as to repel the pests.

(iii) It should not be phytotoxic and soluble in water.

(iv) Should be uniformly spread on treated surface of plants and not leave any harmful residue on plant surfaces. There are some important stomach poisons e.g. lead arsenate, calcium arsenate, methoxychlor, paris green and certain phosphates, BHC, DDT and fluorine compounds.

(i) Lead Arsenates: Lead arsenates were discovered in 1892 in the USA and first used I as insecticide in Massachusetts to control gypsy moths. It acts as stomach poison. lead arsenate exist in two form i.e. in acidic and basic forms. The acid ortho-arsenate (Pb HASO4) or white arsenate burns the tender parts of the plants hence used on plants such as orchards, forests shade trees, shrubs, etc. or the control of chewing insects which are not much important and the basic orthoarsenate [Pb4 (PbOH) (ASO4)3].

It is more stable and less toxic to insects but it needs to be used in greater amounts and takes a longer time compared to the acidic compounds. The commercially available products are mixtures of these two forms. The acid lead arsenate contains about 20% metallic arsenic while the basic arsenate contains 14% metallic arsenic equivalent. It is readily act on digestive enzymes leading death of insects.

Lead arsenate is a relatively weak poison and shows high degree of safety to foliage plants. The acid form decomposes in alkalies, some soaps and hard alkaline waters. lead arsenate may be applied as a dust or spray. It also used as poison bait after mixing with bran and against beetles damaging potatoes and other vegetables.

In India, it was used as spray for the control of lemon butterfly, caster semilooper, Epilachna and Aulacophora beetles on cucurbits and in poison baits for the control of grasshoppers and locusts.

Arsenic toxicity results in the disintegration of the midgut epithelial cells with vacuolated cytoplasm and clumped chromatin of the nuclei. The symptoms of toxication by arsenicals are regurgitation, dizziness and quiescence. Other important arsenicals are calcium arsenate, arsenious oxide, arsenic oxide and paris green.

(ii) Paris green (Emerald green, French green, Mitis green and Schweenfurter green): This is a copper acetoarsenate ECu (C2H302) — 3 Cu (ASO2)2] used first in the U.S.A. in 1865 as a stomach poison for the control of Colorado potato beetle. It is a brilliant green powder, contains 33-39% metallic arsenic, 2-3% of which is water soluble. It is very expensive due to high content of copper, and shows high phototoxic properties as it burns foliage and hence used as mosquito larvicide in limited. It is mostly used in poison baits against pests having chewing types of mouth parts. (i.e. 1-2 kg parisgreen + 25 kg wheat bran + 120 kg of jaggery or molasses). It is mostly used to control leather jackets and slugs. In

India it is not in use. Parisgreen is also one of the constituents of the safe Bordeaux mixture. But the mixing of lime sulphur produces compounds injurious to plants due to which it has been replaced by the recent synthetic organic insecticides.

2. Contact poisons:

The toxic chemical which kills the insects by simple contact or touch is termed as contact poison. The contact poisons are usually utilized as spraying or dusting form. These are act by direct penetration through parts of body wall i.e. through sutures, bases of setae, membranes and through cuticle into the blood (Hemolymph) acting as general or nerve poison. The contact poisons are highly lipophilic so readily absorbed by the liquid present in the epicuticle of exoskeleton.

Contact poisons are naturally occurring compounds like nicotine, pyrethrum and rotenone or synthetic compounds like BHC, DDT and parathion or mixture of compounds from either one or both groups. Generally, these are soluble in plant and insect oils or animal fats. The lethal action of these compounds in the field are said to be due to the following three mechanisms.

(i) Transport from outside the cuticle to the site of action.

(ii) These enters inside the body and usually inhibits the enzymes or other protein interactions.

(iii) Subsequently, acts on nervous, respiratory or other system involving the biochemical effects on the pests.

The contact poisons may be instantaneous in their action when they are called "knockdown poisons" like pyrethrum or may take several hours to kill the pest. Many contact poisons like DDT, BHC are also act stomach poisons if taken into the stomach of an insect. Some of the important contact poisons are BHC, DDT, nicotine preparations, lime sulphur, toxaphene, chiordan, dieldrin, aidrin, methoxychlor, pyrethrum, oil emulsions, rotenone, thiocyanate, TEPP, parathion and systox.

(i) BHC: (Gemmexane, $C_6H_6C_{16}$) Benzene hexachioride (also called 666 and Gemmexane, a trade name given by ICI Ltd. India) was first synthesized in 1825 by Michael Faradey with the name BHC also called HCH (Hexachioro-Cyclo-Hexane). It was discovered as active insecticide by A. P. W. Dupire in France (1941) and by F.D. Leicester in England (1942) independently. But the technical BHC is y — isomer was made known by F. J.D. Thomas in 1943 in England.

Crude BHC is greyish or brownish amorphous solid and has a strong musty (smelling moldy) odor. It is formulated generally as wettable powder or as dust.

BHC (HCH):
It is produced by chlorinating benzene with six atoms of chlorine in presence of sunlight. Since the aromatic character of the ring disappears.

Like other organochiorines, it is insoluble in organic solvents. BHC is a mixture of 6-isomers alpha (a, 55-70%), Beta (f3, 5-14%), Gamma (y, 10-18%), delta (6, 6-8%), etc. and epilson (r and e, 3-4%). Of these, the Beta (13) and epsilon (c) isomers are inactive, the a, 6 and isomers slightly or moderately active and the y (gamma) isomer is the most active. The gamma isomer was first isolated by Von der Linden a German Chemist in 1912 by breaking crude BHC with methyl alcohol or acetic acid. BHC containing 99 per cent of the gamma isomer is called lindane which is odorless unlike BHC. But it is thousand times more active than BHC, does not accumulate in fat of animals and is safer for plants.

Action of BHC: It acts as a contact as well as stomach poison. It is a nerve poison and creates paralysis leading to death. Pharmacological data strongly suggest that it is a neurotoxic agent. It attacks the entire central nervous system and acts as an inhibitor of the Na±, K± and Mg2, ATP asses and blocks the activity of the sodium-potassium pump that maintains the ionic transport of these ions across the nerve membrane and impulse conduction. It is quicker in action than DDT. It depresses the rate of heart beats and elevates the blood pressure. It causes pathological changes in the midgut epithelium also.

The insecticides chiorodane, heptachlor, dieldrin, endrin and endosuiphon are chlorinated cyclic hydrocarbons and grouped under cyclodiene insecticides. BHC is applied as broad spectrum insecticide for the control of most insect pests particularly insects with chewing, mandibulate etc. mouth parts.

(ii) Malathion (Carbonphos): Malathion ($C_{10} H_{19} O_6 PS_2$) is 0, O-dimethyl-S — (1, 2-dicarbethoxy ethyl) phosphordithioate. Technical grade material is 95-98% pure with an unpleasant odor, light yellow to dark brown liquid with a strong offensive odor. It has very low solubility in water, slightly soluble in mineral oils and soluble in most organic solvents, unstable in an alkaline medium and is a contact and stomach poison. Commonly used for the control of a wide variety of pests. It is also applied for the control of mosquitoes, flies, and bed bugs. It has, a low phytotoxicity and the lowest to mammals as compared to all other organophosphates.

Malathion:

It is formulated as 50% (EC) emulsifiable concentrated aerosol, 5% dust, 25% wettable powder, granule; low volume concentrate formulations and very effective against crop pests such as aphids, plant bugs, caterpillars, household pests and livestock pests (lice, fleas, ticks). It is one of the safest chemical available in market. It's acute oral and dermal LD50 values for rat are 1400-1900 and 4000 mg/kg respectively.

3. Systematic Poisons:

Systematic poisons are toxicants when applied to any part of plant, can absorbed (penetrate) plant tissues and translocated (migrated) from the point of application to other parts of the plants in concentrations, thereby rendering the plants toxic to the pests feeding on them. Most modern insecticides have the property of penetrating plant tissues by combining with a lipoid cuticle but only the systematic insecticides can translocated long distances, making the whole plant insecticidal.

Dr. Schrader and co-workers showed very first that some insecticides possess systemic properties. These chemicals have selective in their action and are fairly less harmful to the beneficial insects.

Systemic poisons are used to control the insects having piercing and sucking type of mouth parts i.e. aphids, plant bugs, caterpillars, etc. Since the poisons when applied to the root, stem, leaves or seeds of plants, it is absorbed and reaches different parts of plants.

According to the decomposition of insecticides inside the plant, Ripper (1952) classified systemic poison into three types as follows.

(a) Stable systemic poisons: These insecticides do not decompose or change inside the plant tissue are called stable systemic e.g. selenium.

(b) Endolytic systemic poisons The insecticides which go unchanged from the plant into the insect body and change into a toxicant inside the latter are called endolytic systemic poisons. e.g. Schradan, dimefox.

(c) Endometatoxic systemic poisons: The insecticides which are partly or wholly metabolized into other toxicants inside the plants themselves are called endometatoxic systemic. e.g. demeton.

(i) Thimet (Phorate, C7H1702PS3): It is 0, 0-diethyl-S (ethyl thiomethyl) dithiophosphate. This is a colorless, clear mobile liquid containing not less than 90% phorate, less soluble in water but miscible in many organic solvents like xylene, carbontetrachloride, alcohols; ethers, esters and vegetable oils, unstable in alkaline and humid conditions. It changes into toxic compounds (suiphoxides and sulphones) inside the plants.

Thimet:

It acts as systemic poison, contact poison and fumigant on ins&ts besides this it is also effective against nematodes and mites. It is extensively used for the control of sorghum soot fly, aphids, sucking insects of cotton, tobacco and vegetables, gall midge of rice, etc. Formulated as emulsifiable concentrate, dust and granules. It's acute and dermal LD50 values for rat are 2-3 and 70-300 mg/kg respectively. Thimete is trade name in market.

(ii) Metasystox (Methyldemeton): It is a 0-O-dimethyl-S-32— (ethylthio)-ethyl phosphorothiolate belongs to organophosphorus. The technical grade product is a mixture of thiono and thiolo isomers of 0, 0-dimethyl-, 1 -2- ethylmercapto-ethyl-thiophosphate. The product contains a small amount of trimethylthiophosphate, 2- hydroxydiethylsuiphide and some other ester of of thisphosphoric acid occur as impurities. The thiono isomer is a liquid with a characteristic, unpleasant odor and B.P. 930 C, slightly soluble in water. While thioloisomer is a liquid and boiling at 102°C. It is more soluble in water than the thionosomer. Both the isomers are highly soluble in organic solvents. It is formulated as 25% and 50% effective emulsified concentrate.

Metasystox after entering the plant body is readily Isomerised to the thiolo isomer which is further metabolised. Being a organo phosphorous compound, it is irreversible inhibitor of the cholinesterase enzymes of the neuromuscular system.

4. Fumigants:

Fumigants are the chemicals or insecticides in a gaseous form that enter the body through the spiracles and spread all over via the tracheal system of the pests (insects). Fumigants kill insects, in two ways.

(i) By clogging the tracheal system and cutting it off from the oxygen supply.

(ii) By their own toxic action interfering with the respiratory process. Therefore, these chemicals, kill the insects irrespective of their mouth parts or feeding habits.

Fumigants are used in vapor state to keep the insect population under check in buildings, storage, ship holds and sometimes in the soil. Fumigants are commonly mixture of two or more gases formulated in form of liquid under pressure e.g. phosphene or hydrogen phosphide gas is released from the tables containing aluminium phosphide and ammonium carbonate in the presence of moisture.

A good fumigant should possess:
(i) A good volatile with a deep penetrating power.
(ii) Should not corrosive to the equipments and household articles.
(iii) Should not be inflammable.
(iv) It should be not harm to man and plants.

Generally, fumigants are used in warehouses, kothis, bins, airtight containers, tent houses, ships, airplanes, factories, mills, hotels, prisons, theaters, etc. to kill insects like ants, termites, grubs, nematodes, stem borers, bark eating caterpillars, intestinal worms, chinch bugs, grasshoppers, etc.

(i) Carbon tetrachioride (CCl_4): This is a colorless liquid or gas with a chloroform like smell (pungent) sp. gr. 1.595 at 20^0 C and vapor pressure 114.5 mm Hg at 20° C; insoluble in water and soluble in organic solvents. Not inflammable or explosive. It has low insect toxicity. Commonly it is used alone only where fire hazards are acute or in small scale fumigation. But when used with carbon disuiphide Or ethylene dibromide it reduces fire hazards. It is also used as diluent to increase the volatility and distribution of other fumigants like methyl bromide, ethylene dichloride and chioropicrin. Carbon tetrachioride gas enters the insect spiracles during respiration leading to clogging and finally death of insects. Carbon tetrachioride is very toxic to warm blooded animals causing severe kidney and liver damage.

(ii) Ethylene dichloride ($C_2H_4Cl_2$).: Ethylene dichioride first synthesized by cotton and roark in 1927. It is a colorless sweet smelling liquid with B.P. 83.5°C, M.P. — 36°C and sp. gr. 1.25, miscible with organic solvents. It can remain stable with water, alkalis and acids. It is highly inflammable with low mammalian toxicity and moderate insect toxicity and mainly used to control the all types of stored grains.

(iii) Ethylene dichioride is commonly mixed with the carbon tetrachioride and available in the market as a EDCT mixture of 3:1 proportion (EDCT). EDCT is harmless to seeds, less toxic to man and other mammals. EDCT is available in the form of liquid which is volatile at the room temperature and so easy in handling as it can be applied by pouring it over clothes or on grain bags or grain bins. Emulsion of ethylene dichloride prepared in water is used to control the peach tree borer. Food substances rich in fats absorb appreciable amounts of the ethylene dichloride gas and requires long periods of aeration to remove the taint and smell.

CULTURAL CONTROL

Definition: Manipulation of cultural practices to the disadvantage of pests.

I. Farm level practices

Croping Techniques	Pest checked
Ploughing	Red hairy caterpillar
Puddling	Rice Meal bug
Trimming and plastering	Rice grass Hooper
Pest free seed material	Potato tuber moth
High seed rate	Sorghum shoofly
Rouge space planting	Rice brown plant Hooper
Plant density	Rice brown plant Hooper
Earthing up	Sugarcane whitefly
Detrashing	Sugarcane whitefly
Destruction of weed host	Citrus fruit sucking moth
Destruction of alternative host	Cotton white fly
Flooding	Rice army worm
Trash mulching	Sugarcane early shoot borer
Pruning / topping	Rice stem borer
Intercropping	Sorghum stem borer
Trap cropping	Diamond hock moth
Water management	Brown plant Hooper
Judicial applications of fertilizers	Rice leaf folder
Timely harvesting	Sweet potato weevil

II. Community level practices

1. Synchronized sowing: Dilution of pest infestation (eg) Rice, Cotton
2. Crop rotation: Breaks insect life cycle
3. Crop sanitation
    a) Destruction of insect infested parts (eg.) Mealy bug in brinjal
    b) Removal of fallen plant parts (eg.) Cotton squares
    c) Crop residue destruction (eg.) Cotton stem weevil

Advantages

1. No extra skill
2. No costly inputs
3. No special equipments
4. Minimal cost
5. Good component in IPM
6. Ecologically sound

Disadvantages

1. No complete control
2. Prophylactic nature
3. Timing decides success

PHYSICAL CONTROL

Modification of physical factors in the environment to minimize (or) prevent pest problems. Use of physical forces like temperature, moisture, etc. in managing the insect pests.

A. Manipulation of temperature

1. Sun drying the seeds to kill the eggs of stored product pests.
2. Hot water treatment (50 - 55°C for 15 min) against rice white tip nematode.
3. Flame throwers against locusts.
4. Burning torch against hairy caterpillars.
5. Cold storage of fruits and vegetables to kill fruit flies (1 - 2°C for 12 - 20 days).

B. Manipulation of moisture

1. Alternate drying and wetting rice fields against BPH.
2. Drying seeds (below 10% moisture level) affects insect development.
3. Flooding the field for the control of cutworms.

C. Manipulation of light

1. Treating the grains for storage using IR light to kill all stages of insects (eg.) Infra-red seed treatment unit.
2. Providing light in storage go downs as the lighting reduces the fertility of Indian meal moth, Plodia.
3. Light trapping.

D. Manipulation of air

1. Increasing the CO_2 concentration in controlled atmosphere of stored grains to cause asphyxiation in stored product pests.

E. Use of irradiation

Gamma irradiation from $Co60$ is used to sterilize the insects in laboratory which compete with the fertile males for mating when released in natural condition. (eg.) cattle screw worm fly, Cochliomyia hominivorax control in Curacao Island by E.F. Knipling.

F. Use of greasing material

Treating the stored grains particularly pulses with vegetable oils to prevent the ovi-position and the egg hatching. e.g., bruchid adults.

G. Use of visible radiation: Yellow colour preferred by aphids, cotton whitefly: yellow sticky traps.

H. Use of Abrasive dusts

1. Red earth treatment to red gram: Injury to the insect wax layer.
2. Activated clay: Injury to the wax layer resulting in loss of moisture leading to death. It is used against stored product pests.
3. Drie-Die: This is a porous finely divided silica gel used against storage insects.

MECHANICAL CONTROL

Use of mechanical devices or manual forces for destruction or exclusion of pests:

A. Mechanical destruction:
Life stages are killed by manual (or) mechanical force.

Manual Force

1. Hand picking the caterpillars
2. Beating: Swatting housefly and mosquito
3. Sieving and winnowing: Red flour beetle (sieving) rice weevil (winnowing)
4. Shaking the plants: Passing rope across rice field to dislodge caseworm and shaking neem tree to dislodge June beetles
5. Hooking: Iron hook is used against adult rhinoceros beetle
6. Crushing: Bed bugs and lice
7. Combing: Delousing method for Head louse
8. Brushing: Woolen fabrics for clothes moth, carper beetle.

Mechanical force

1. Entoleter: Centrifugal force - breaks infested kernels - kill insect stages - whole grains unaffected - storage pests.
2. Hopper dozer: Kill nymphs of locusts by hording into trenches and filled with soil.
3. Tillage implements: Soil borne insects, red hairy caterpillar.
4. Mechanical traps: Rat traps of various shapes like box trap, back break trap, wonder trap, Tanjore bow trap.

B. Mechanical exclusion:
Mechanical barriers prevent access of pests to hosts.

1. Wrapping the fruits: Covering with polythene bag against pomegranate fruit borer.
2. Banding: Banding with grease or polythene sheets - Mango mealy bug.
3. Netting: Mosquitoes, vector control in green house.
4. Trenching: Trapping marching larvae of red hairy caterpillar.
5. Sand barrier: Protecting stored grains with a layer of sand on the top.
6. Water barrier: Ant pans for ant control.
7. Tin barrier: Coconut trees protected with tin band to prevent rat damage.
8. Electric fencing: Low voltage electric fences against rats.

Advantages of mechanical control

1. Home labor utilization
2. Low equipment cost
3. Ecologically safe

Disadvantages

1. Rarely highly effective
2. Labor intensive

Mechanical Appliances in controlling the pests:

1. Light traps: Most adult insects are attracted towards light in night. This principle is used to attract the insect and trapped in a mechanical device.

a) Incandescent light trap: They produce radiation by heating a tungsten filament. The spectrum of lamp include a small amount of ultraviolet, considerable visible especially rich in yellow and red. Simple incandescent light trap, portable incandescent electric. Place a pan of kerosenated water below the light source.

b) Mercury vapor lamp light trap: They produce primarily ultraviolet, blue and green radiation with little red. (eg.) Robinson trap. This trap is the basic model designed by Robinson in 1952. This is currently used towards a wide range of Noctuids and other nocturnal flying insects. A mercury lamp (125 W) is fixed at the top of a funnel shaped (or) trapezoid galvanized iron cone terminating in a collection jar containing dichlorvos soaked in cotton as insecticide to kill the insect.

c) Black light trap: Black light is popular name for ultraviolet radiant energy with the range of wavelengths from 320-380 nm. Some commercial type like Pest-O-Flash, Keet-O-Flash are available in market. Flying insects are usually attracted and when they come in contact with electric grids, they become electrocuted and killed.

2. Pheromone trap: Synthetic sex pheromones are placed in traps to attract males. The rubberized septa, containing the pheromone lure are kept in traps designed specially for this purpose and used in insect monitoring / mass trapping programmes. Sticky trap, water pan trap and funnel type models are available for use in pheromone based insect control programmes.

3. Yellow sticky trap: Cotton whitefly, aphids, thrips prefer yellow colour. Yellow colour is painted on tin boxes and sticky material like castor oil / Vaseline is smeared on the surface. These insects are attracted to yellow colour and trapped on the sticky material.

4. Bait trap: Attractants placed in traps are used to attract the insect and kill them. (eg.) Fishmeal trap: This trap is used against sorghum shoot fly. Moistened fish meal is kept in polythene bag or plastic container inside the tin along with cotton soaked with insecticide (DDVP) to kill the attracted flies

5. Pitfall trap helps to trap insects moving about on the soil surface, such as ground beetles, collembolan, spiders. These can be made by sinking glass jars(or) metal cans into the soil. It consists of a plastic funnel, opening into a plastic beaker containing kerosene supported inside a plastic jar.

6. Probe trap: Probe trap is used by keeping them under grain surface to trap stored product insect.

7. Emergence trap: The adults of many insects which pupate in the soil can be trapped by using suitable covers over the ground. A wooden frame covered with wire mesh covering and shaped like a house roof is placed on soil surface. Emerging insects are collected in a plastic beaker fixed at the top of the frame.

8. Indicator device for pulse beetle detection: A new cup shaped indicator device has been recently designed to predict timely occurrence of pulse beetle *Callosobruchus* spp. This will help the farmers to know the correct time of emergence of pulse beetle. This will help them in timely sun drying which can bill all the eggs.

Biological control
Definition
The study and utilization of parasitoids, predators and pathogens for the regulation of pest population densities.
Biological control can also be defined as the utilization of natural enemies to reduce the damage caused by noxious organisms to tolerable levels. Biological control is often shortened to biocontrol.

Factors affecting biological control

1. Tolerance limit of crop to insect injury - Successful in crops with high tolerance limit
2. Crop value - Successful in crops with high economic value
3. Crop duration - Long duration crops highly suitable
4. Indigenous or Exotic pest - Imported NE more effective against introduced pest
5. If alternate host available for NE, control of target pest is less
6. If unfavourable season occurs, reintroduction of NE required
7. Presence of hyperparasites reduces effectiveness of biocontrol
8. Tritrophic interaction of Plant-Pest-Natural enemy affects success of biocontrol, e.g. Helicoverpa parasitization by Trichogramma more in timato than corn
9. Use of pesticides affect natural enemies
10. Selective insecticides (less toxic to NE required)
11. Identical situation for successful control does not occur

Qualities of an effective natural enemy

1. Adaptable to the environmental condition
2. Host specific (or narrow host range)
3. Multiply faster than the host (with high fecundity)
4. Short life cycle and high female: male ratio
5. High host searching capacity
6. Amenable for easy culturing in laboratory
7. Dispersal capacity
8. Free from hyper parasites
9. Synchronize life cycle with host

THREE MAJOR TECHNIQUES OF BIOLOGICAL CONTROL

1. Conservation and encouragement of indigenous Natural Enemies

Defined as actions that preserve and increase Natural Enemies by environmental manipulation. e.g. Use of selective insecticides, provide alternate host and refuge for Natural Enemies.

2. Importation or Introduction

Importing or introducing Natural Enemies into a new locality (mainly to control introduced pests).

3. Augmentation

Propagation (mass culturing) and release of Natural Enemies to increase its population.

Two types,

(i) Inoculative release: Control expected from the progeny and subsequent generations only.

(ii) Inundative release: NE mass cultured and released to suppress pest directly, e.g. Trichogramma sp. egg parasitoid, Chrysoperla carnia predator

ROLE OF PARASITOIDS AND PREDATORS IN IPM

- Parasitoids and predators may be used in Agriculture and IPM in three ways. They are:

i) Conservation

ii) Introduction

iii) Augmentation - (a) Inoculative release, (b) Inundative release

- Since biological control is safe to environment, it should be adopted as an important component of IPM.
- Biological control method can be integrated well with other methods namely cultural, chemical methods and host plant resistance (except use of broad spectrum insecticides)
- Biological control is self propagating and self perpetuating
- Pest resistance to Natural Enemies is not known
- No harmful effects on humans, livestock and other organisms
- Biological control is virtually permanent
- Biological agents search and kills the target pest

MICROBIAL CONTROL

- It is a branch of biological control
- Defined as control of pests by use of microorganisms like viruses, bacteria, protozoa, fungi, rickettsia and nematodes.

I. VIRUSES

Viruses coming under family Baculoviridae cause disease in lepidoptera larvae. Two types of viruses are common.

NPV (Nucleopolyhedro virus) e.g. HaNPV, SlNPV

GV (Granulovirus) e.g. CiGV

Symptoms

Lepidopteron larva become sluggish, pinkish in colour, lose appetite, body becomes fragile and rupture to release polyhedra (virus occlusion bodies). Dead larva hang from top of plant with prolegs attached (Tree top disease or "Wipfelkrankeit")

II. BACTERIA

Spore forming (Facultative - Crystalliferous)

2 types of bacteria

Spore forming (Obligate) & Non spore forming

i. Spore forming (Facultative, Crystalliferous)

The produce spores and also toxin (endotoxin). The endotoxin paralyses gut when ingested e.g. Bacillus thuringiensis effective against lepidopteran. Commercial products - Delfin, Dipel, Thuricide

ii. Spore-forming (Obligate)

e.g. Bacillus popilliae attacking beetles, produce 'milky disease'

Commercial product - 'Doom' against 'white grubs'

iii. Non-spore forming

e.g. Serratia entomophila on grubs

III. FUNGI

i. Green muscardine fungus - Metarhizium anisopliae attack coconut rhinoceros beetle
ii. White muscardine fungus - Beaveria bassiana against lepidopteran larvae
iii. White halo fungus - Verticillium lecanii on coffee green scale.

Other Microbes: Protozoa, Nematodes

Limitations of biocontrol technique

- Complete control not achieved - Slow process
- Subsequent pesticide use restricted
- Expensive to culture many NE
- Requires trained man power

PHEROMONES

Semiochemicals are chemical substances that mediate communication between organisms. Semiochemicals maybe classified into Pheromones (intraspecific semiochemicals) and Allelochemics (interspecific semiochemicals).

Pheromones are chemicals secreted into the external environment by an animal which elicit a specific reaction in a receiving individual of the same species. Pheromones are volatile in nature and they aid in communication among insects.

Pheromones are exocrine in origin (i.e. secreted outside the body). Hence they were earlier called as ectohormones. In 1959, German chemists Karlson and Butenandt isolated and identified the first pheromone, a sex attractant from silkworm moths. They coined the term pheromone. Since this first report, hundreds of pheromones have been identified in many organisms. The advancement made in analytical chemistry aided pheromone research.

Based on the responses elicited pheromones can be classified into 2 groups:

a) Primer pheromones: They trigger off a chain of physiological changes in the recipient without any immediate change in the behaviour. They act through gustatory (taste) sensilla. (e.g.) Caste determination and reproduction in social insects like ants, bees, wasps, and termites are mediated by primer pheromones. These pheromones are not of much practical value in IPM.

b) Releaser pheromones: These pheromones produce an immediate change in the behaviour of the recipient. Releaser pheromones may be further subdivided based on their biological activity into

1) Sex pheromones
2) Aggregation pheromones
3) Alarm pheromones
4) Trail pheromones

Releaser pheromones act through olfactory (smell) sensilla and directly act on the central nervous system of the recipient and modify their behaviour. They can be successfully used in pest management programmes.

1) Sex pheromones are released by one sex only and trigger behaviour patterns in the other sex that facilitate in mating. They are most commonly released by females but may be released by males also. In over 150 species of insects, females have been found to release sex pheromones and about 50 species males produce.

Aphrodisiacs are substances that aid in courtship of the insects after the two sexes are brought together. In many cases males produce aphrodisiacs.

Table 5: Properties of sex pheromone

Properties	Female sex pheromone	Male sex pheromone
Range	Act at a long range. Attracts males from long distance.	Acts at a short distance
Role of other stimuli	Play less role	Visual and auditory stimuli play major role
Action elicited in the other sex	Attracts and excites males to copulate	Lowers females resistance to mating
Importance in IPM	More important	Less important

Insect orders producing sex pheromones

Lepidoptera, Orthoptera, Dictyoptera, Diptera, Coleoptera, Hymenoptera, Hemiptera, Neuroptera and mecoptera. In Lepidoptera, sex pheromonal system is highly evolved.

Pheromone producing glands

In Lepidoptera they are produced by eversible glands at the tip of the abdomen of the females. The posture shown during pheromone release is called 'calling position'. Aphrodisiac glands of male insects are present as scent brushes (or hair pencils) at the tip of the abdomen (eg. Male butterfly of Danaus sp.). Andraconia are glandular scales on wings of male moths producing aphrodisiacs.

Pheromone reception

Female sex pheromones are usually received by olfactory sensillae on male antennae and males search upwind, following the odour corridor of the females. In pheromone perceiving insects, the antennae of male moths are larger and greatly branched than female moths to accommodate numerous olfactory sensilla.

Chemical nature of sex pheromones

In general pheromones have a large number of carbon atoms (10-20) and high molecular weight (180 – 300 daltons). Narrow specificity and high potency are two attributes which depend on long chain carbon atoms and high molecular weight. But since pheromones are volatile their molecular weights cannot be very high as they cannot be carried by wind.

Name	Biological name	Insect pheromone
Silkworm	Bombyx mori	Bombykol
Gypsy moth	Porthesia dispar	Gyplure, disparlure
Pink bollworm	Pectinophora gossypiella	Gossyplure
Cabbage looper	Trichoplusia ni	Looplure
Tobacco cutworm	Spodoptera litura	Spodolure, litlure
Gram pod borer	Helicoverpa armigera	Helilure
Honey bee queen	Apis sp	9-oxy desonoic acid

Butenandt and his coworkers in 1959 isolated 12mg of pheromone from the abdomen of half a million virgin females of silkworm. They named the pheromone as Bombykol. The chemical name is 10,12 – hexadeca dienol. It is a primary alcohol.

The following are some of the female sex pheromones identified in insects:

Examples of male sex pheromones:
Cotton boll weevil, *Anthonomas grandis*, Coleoptera
Cabbage looper, *Trichoplusia ni*, Lepidoptera
Mediterranean fruitfly, *Ceratitis capitata*, Diptera.
Multi-component pheromone system: If the pheromone of an insect is composed of only one chemical compound we call it mono-component pheromone system.
Pheromones of some insects contain more than one chemical compound. In this case we call it as multi-component pheromone system. The sex pheromone of two different species may contain same chemical compounds but the ratio of the compounds may vary. This brings about species specificity.

Pest Management With Sex Pheromones:
Synthetic analogues of sex pheromones of quite large No. of pests are now available for use in Pest management. Sex pheromones are being used in pest management in three different ways.
a) In sampling and detection (Monitoring)
b) To attract and kill (Mass trapping)
c) To disrupt mating (Confusion or Decoy method)

a) In sampling and detection (Monitoring):

Pheromones can be used for monitoring pest incidence/ outbreak in the following ways:

1. Sterility
2. Insect attractants
3. Insect repellents
4. Antifeedants
5. Insect growth regulators

STERILITY METHODS

Definition

Control of pest population achieved by releasing large number of sterilised male insects, which will compete with the normal males and reduce the insect population in subsequent generation. It is usually referred as SIT (Sterile insect technique) or SIRM (Sterile insect release method). Sterile insect release method is a genetic control method. This is also called Autocidal control since insects are used against members of their own species.

E.F. Knipling in 1937 in South East USA used the SIRM technique to control the screw wormfly *(Cochliomyia nominivorax)* a serious livestock pest.

The sterile to fertile male ratio, called S:F ratio is important, as the reduction in reproductive potential of natural population depends on S:F ratio.

The mating with the sterile males will produce inviable or sterile eggs.

Trend of hypothetical population subjected to SIRM Assumption

Female: Male ratio 1:1

1 female produces 5 females as off spring in one generation

Table 6 showing the number of sterile individuals required:

Generation	No.of females without releases	No.of sterile males released	No.of females releases(9:1)	Ratio sterile to normal males	No. of fertile females
1.	1,000,000	9,000,000	1,000,000	9:1	100,000
2.	5,000,000	9,000,000	500,000	18:1	26,316
3.	25,000,000	9,000,000	131,579	68:1	1,907
4.	125,000,000	9,000,000	9,535	944:1	10
5.	625,000,000	9,000,000	50	180,000:1	0

In suitable circumstances sterile male release method (SIRM) can be more effective, compared to insecticide application.

Generation	No. of females with no treatment	No. of females with sterile release (9:1)	No. of females with insecticide (90% kill)
1.	1,000,000	1,000,000	1,000,000
2.	5,000,000	500,000	500,000
3.	25,000,000	131,579	250,000
4.	125,000,000	9,535	125,000
5.	625,000,000	50	62,500
6.	3,125,000,000	0	31,250

SIRM technique can also be used after insecticide application which will be more effective. Circumstances for using this method:

- Against well established pest when their population density is low
- Against newly introduced pest
- Against isolated population as in island
- Combined with cultural and chemical methods

Methods of sterilization

Chemosterilants: Any chemical which interfere with the reproductive capacity of an insect. Alkylating agents: They inhibit nucleic acid synthesis inhibit gonad development produce mutagenic effect (e.g. TEPA, Chloro ethylamine)

Antimetabolites: Chemicals having structural similarity to biologically active substances. They interfere with nucleic acid synthesis. e.g. 5-Fluororacil, Amithopterin

Irradiation: Irradiation done by exposing insects to , , radiations, X rays and neutrons. Of these, -radiation by ^{60}CO (cobalt) with its half-life of 60 years is the most common method. Irradiation causes following sterility effects in insects

Infecundity

Aspermia

Inability to mate

Dominant lethal mutation

Radiation dose required for different species and stages for sterilization (expressed as rads - radiation absorbed dose).

Insect Stage	Dose
Housefly 2-3 day pupae	3000 rads
Screw worm 5 day pupae	2500 rads
1 day adult	5000 rads

Sterilizing natural population: In this method, instead of releasing sterilised males into the field, a chemosterilant is sprayed in field like insecticide. The chemosterilant sterilizes both male and female. These do not produce offspring-equivalent to killing them.

Bonus effect: The bonus effect of this method is that the sterilized males mate with normal females and reduce their reproductive capacity.

Chemosterilants used are TEPA, HEMPA, BISULFAN, etc.

Requirements for SIRM

A method inducing sterility without impairing sexual behaviour of insects.

Mass rearing of the insects

Information on population density and its rate of increase

The released insects must not cause damage to the crops, livestock or human beings

Good intermingling of released and natural population

Releasing sterilized insects when the wild population is abundant

This method is effective against newly introduced pest or isolated insect population as in island.

There should be high sterile to fertile (S:F) ratio for quicker control.

Limitations of SIRM

Not effective against insects which are prolific breeders

Sterilizing and mutagenic effect of chemosterilants and irradiation cause problem in higher animals and man (Carcinogenic and mutagenic).

Insect Growth Regulators (IGRs) are compounds which interfere with the growth, development and metamorphosis of insects. IGRs include synthetic analogues of insect hormones such as ecdysoids and juvenoids and non-hormonal compounds such as precocenes (Anti JH) and chitin synthesis inhibitors.

Natural hormones of insects which play a role in growth and development are:

Brain hormones: The are also so called activation hormones (AH). AH is secreted by neuro secretory cells (NSC) which are neurons of central nervous system (CNS). It's role is to activate the corpora allata to produce juvenile hormone (JH).

Juvenile hormone (JH): Also called neotinin. It is secreted by corpora allata which are paired glands present behind insect brain. Their role is to keep the larva in juvenile condition. JH I, JH II, JH III and JH IV have been identified in different groups of insects. The concentration of JH decreases as the larva grows and reaches pupal stage. JH I, II and IV are found in larva while JH III is found in adult insects and are important for development of ovary in adult females.

Ecdysone: Also called Moulting hormone (MH). Ecdysone is a steroid and is secreted by Prothoracic Glands (PTG) present near prothoracic spiracles. Moulting in insects is brought about only in the presence of ecdysone. Ecdysone level decreases and is altogether absent in adult insects.

IGRs used in Pest management

Ecdysoids: These compunds are synthetic analogues of natural ecdysone. When applied in insects, kill them by formation of defective cuticle. The development processes are accelerated bypassing several normal events resulting in integument lacking scales or wax layer.

Juvenoids (JH mimics): They are synthetic analogues of Juvenile Hormone (JH). They are most promising as hormonal insecticides. JH mimics were first identified by Williams and Slama in the year 1966. They found that the paper towel kept in a glass jar used for rearing a *Pyrrhocoris* bug caused the bug to die before reaching adult stage. They named the factor from the paper as 'paper factor' or 'juvabione'. They found that the paper was manufactured from the wood pulp of balsam fir tree *(Abies balsamea)* which contained the JH mimic.

Juvenoids have anti-metamorphic effect on immature stages of insect. They retain *status quo* in insects (larva remains larva) and extra (super numerary) moultings take place producing super larva, larval-pupal and pupal-adult intermediates which cause death of insects. Juvenoids are larvicidal and ovicidal in action and they disrupt diapause and inhibit embryogenesis in insects.

Methoprene is a JH mimic and is useful in the control of larva of hornfly, stored tobacco pests, green house homopterans, red ants, leaf mining flies of vegetables and flowers

Anti JH or Precocenes: they act by destroying corpora allata and preventing JH synthesis.

When treated on immature stages of insect, they skip one or two larval instars and turn into tiny precocious adults. They can neither mate, nor oviposit and die soon. Eg. EMD, FMev, and PB (Piperonyl Butoxide)

Chitin Synthesis inhibitors: Benzoyl phenyl ureas have been found to have the ability of inhibiting chitin synthesis in vivo by blocking the activity of the enzyme chitin synthetase. Two important compounds in this category are Diflubenzuron (Dimilin) and Penfluron. The effects they produce on insects include

Disruption of moulting Displacement of mandibles and labrum Adult fails to escape from pupal skin and dies Ovicidal effect.

Chitin sysnthesis inhibitors have been registered for use in many countries and used successfully against pests of soybean, cotton, apple, fruits, vegetables, forest trees and mosquitoes and pests of stored grain

IGRS from Neem: Leaf and seed extracts of neem which contains azadirachtin as the active ingredient, when applied topically causes growth inhibition, malformation, mortality and reduced fecundity in insects.

Hormone mimics from other living organisms: Ecdysoids from plants (Phytoecdysones) have been reported from plants like mulberry, ferns and conifers. Juvenoids have been reported from yeast, fungi, bacteria, protozoans, higher animals and plants.

Advantages of Using IGRs

Effective in minute quantities and so are economical Target specific and so safe to natural enemies Bio-degradable, non-persistent and non-polluting Non-toxic to humans, animals and plants

Disadvantages

Kills only certain stages of pest Slow mode of action

Since they are chemicals possibility of build-up of resistance Unstable in the environment

ANTIFEEDANTS

Antifeedants are chemicals that inhibit feeding in insects when applied on the foliage (food) without impairing their appetite and gustatory receptors or driving (repelling) them away from the food. They are also called gustatory repellents, feeding deterrents and rejectants. Since do not feed on trated surface they die due to starvation.

Groups of antifeedants

Triazenes: AC 24055 has been the most widely used triazene which is a oduorless, tasteless, non-toxic chemical which inhibit feeding in chewing insects like caterpillars, cockroaches and beetles.

Organotins. They are compounds containing tin. Triphenyl tin acetate is an important antifeedants in this group effective against cotton leaf worm, Colarado potato beetle, caterpillars and grass hoppers

Carbamates: At sublethal doses thiocarbamates and phenyl carbamates act as antifeedants of leaf feeding insects like caterpillars and Colarado potato beetle. Baygon is a systemic antifeedants against cotton boll weevil.

Botanicals: Antifeedants from non-host plants of the pest can be used for their control The following antifeedants are produced from plants.

Pyrethrum: Extracted from flowers of *Chrysanthemum cinerarifolium* acts as antifeedants at low doses against biting fly, Glossina sp.

Neem: Extracted from leaves and fruits of neem *(Azadirachta indica)* is an antifeedant against many chewing pests and desert locust in particular Apple factor: Phlorizin is extracted from apple which is effective against nonapple feeding aphids.

Solanum alkaloids: Leptine, tomatine and solanine are alkaloids extracted from Solanum plants and are antifeedants to leaf hoppers.

Miscellaneous compounds: Compounds like copper stearate, copper resinate, mercuric chloride and Phosphon are good antifeedants.

Mode of action: Antifeedants inhibit the gustatory (taste) receptors of the mouth region. Lacking the right gustatory stimulus the insect fails to recognize the trated leaf as food. The insect slowly dies due to starvation.

Advantages

Affect plant feeders, but safe to natural enemies Pest not immediately killed, so natural enemies can feed on them No phytotoxicity or pollution Disadvantages

Only chewing insects killed and not sucking insects

Not effective as sole control measure, can be included in IPM

INSECT ATTRACTANTS

Chemicals that cause insects to make oriented movements towards their source are called insect attractants. They influence both gustatory (taste) and olfactory (smell) receptors.

Types of Attractants

Pheromones: Pheromones are chemicals secreted into the external environment by an animal which elicit a specific reaction in a receiving individual of the same species.

Food lures: Chemical present in plants that attract insect for feeding. They stimulate olfactory receptors.

List of natural and synthetic food lures

Insects	Lure
	Natural
Pests of cruciferae	Isothiocyanates from seeds of cruciferae
Onion fly (Hylemya antiqua)	Propylmercaptan from onions
Bark beetle	Terpenes from barks
Housefly	Sugar and molasses
	Synthetic
Oriental fruitfly (Dacus dorsalis)	Methyl eugenol
Melon fruitfly (Dacus cucurbitae)	Cuelure
Mediterranean fruitfly (Ceratitis capitata)	Trimedlure

Oviposition lures: These are chemicals that govern the selection of suitable sites for oviposition by insects. For example extracts of corn attracts *Helicoverpa armigera* for egg laying on any treated surface.

Use of Attractants in IPM Insect attractants are used in 3 ways in pest management

a) Sampling and monitoring pest population

b) Luring pests to insecticide coated traps or poison baits

Examples of poison baits

For biting insects: Moistened Bran + molasses) + insecticides

For sucking insects: Sugar solution + insecticide

For fruitflies: Trimedlure/ Cuelure/ Methyl eugenol + insecticides

For cockroaches: Sweet syrup + white or yellow phosphorus

For sweet-loving ants: Thallous sulphste + sugar + honey + glycerine + water

For meat loving ants: Thallous sulphate + peanut butter

c) in distracting insects from normal mating, aggregation, feeding or oviposition

The female insects, if lured to wrong plants for egg laying, the emerging larva will starve to death.

Advantage of using attractants is that they are specific to target insects and NE not affected. But they cannot be relied as the sole method of control and can only be included in IPM as a component.

INSECT REPELLENTS

Chemicals that induce avoiding (oriented) movements in insects away from their source are called repellents. They prevent insect damage to plants or animals by rendering them unattractive, unpalatable or offensive.

Types of repellents

Physical repellents: Produce repellence by physical means

Contact stimuli repellents: Substances like wax or oil when applied on leaf surface changes physical texture of leaf which are disagreeable to insects

Auditory repellents: Amplified sound is helpful in repelling mosquitoes.

Barrier repellents: Tar bands on trees and mosquito nets are examples.

Visual repellents: Yellow light acts as visual repellents to some insects.

Feeding repellents: Antifeedants are feeding repellents. They inhibit feeding.

Chemical repellents: Repellents of Plant origin: Essentials oils of Citronella, Camphor and cedarwood act as repellents. Commercial mosquito repellent 'Odomos' uses citronella oil extracted from lemongrass, Andrpogon pardus as repellent.

Pyrethrum extracted form Chrysanthemum is a good repellent and has been used against tsetse fly, *Glossina morsitans*.

Synthetic repellents: Repellents synthetically produced.

List of important synthetic repellents

Insects	Repellents
Mosquito, blood suckers	Dimethyl pthalate
Mites (chiggers)	Benzyl benzoate
Crawling insects	Trichlorobenzene
Phytophagous insects	Bordeaux mixture
Wood feeders	Pentachlorophenol
Fabric eaters	Naphthalene or mothballs
Bees	Smoke

Uses of repellents

They can be applied on body to ward off insects Used as fumigants in enclosed area.

Used as sprays on domestic animals

To drive away insects from their breeding place.

BIORATIONAL CONTROL

Controlling insects using chemicals that affect insect behaviour, growth or reproduction, is called biorational control.

Insect Growth Regulator,

Chitin synthesis inhibitor,

JH analogues, Anti JH,

Moulting hormone,

Pheromones ƒ All these methods are included in

Allelochemics Biorational method of control

Attractant, Repellent,

Antifeedant,

Chemosterilant,

Sterile male release

They are called biorational agents in pest control, because of their selective nature in killing only the target insects without affecting non target organisms.

The desired effect of a pesticide can be obtained only if it si applied by an appropriate method in appropriate time. The method of application depends on nature of pesticide, formulation, pests to be managed, site of application, availability of water etc.

Dusting: Dusting is carried out in the morning hours and during very light air stream. It can be done manually or by using dusters. Some times dust can be applied in soil for the control of soil insects. Dusting is cheaper and suited for dry land crop pest control.

Spraying: Spraying is normally carried out by mixing EC (or) WP formulations in water.

There are three types of spraying.

	Spray fluid (litre per acre)	Droplet size	Area covered per day	Equipment used
a) High volume spraying	200-400	150	2.5 ac	Knapsack, Rocker sprayers
b) Low volume spraying	40-60	70-150	5.6 ac	Power sprayer, Mist blower
c) Ultra low volume spraying	2-4 lit.	20-70	20 ac	ULV sprayer, Electrodyn sprayer

Granular application: Highly toxic pesticides are handled safely in the form of granules. Granules can be applied directly on the soil or in the plant parts. The methods of application are

Broadcasting: Granules are mixed with equal quantity of sand and broadcasted directly on the soil or in thin film of standing water. (eg.) Carbofuran 3G applied @ 1.45 kg/8 cent rice nursery in a thin film of water and impound water for 3 days.

Infurrow application: Granules are applied at the time of sowing in furrows in beds and covered with soil before irrigation. (eg.) Carbofuran 3G applied @ 3 g per meter row for the control of sorghum shootfly.

Side dressing: After the establishment of the plants, the granules are applied a little away from the plant (10-15 cm) in a furrow.

Spot application: Granules are applied @ 5 cm away and 5 cm deep on the sides of plant. This reduces the quantity of insecticide required.

Ring application: Granules are applied in a ring form around the trees.

Root zone application: Granules are encapsulated and placed in the root zone of the plant. (eg.) Carbofuran in rice.

Leaf whorl application: Granules are applied by mixing it with equal quantity of sand in the central whorl of crops like sorghum, maize, sugarcane to control internal borers.

Pralinage: The surface of banana sucker intended for planting is trimmed. The sucker is dipped in wet clay slurry and carbofuran 3G is sprinkled (20-40 g/sucker) to control burrowing nematode.

Seed pelleting/seed dressing: The insecticide mixed with seed before sowing (eg.) sorghum seeds are treated with chlorphyriphos 4 ml/kg in 20 ml of water and shade dried to control shootfly. The carbofuran 50 SP is directly used as dry seed dressing insecticide against sorghum shootfly.

Seedling root dip: It is followed to control early stage pests (eg.) in rice to control sucking pests and stem borer in early transplanted crop, a shallow pit lined with polythene sheet is prepared in the field. To this 0.5 kg urea in 2.5 litre of water and 100 ml chlorpyriphos in 2.5 litre of water prepared separately are poured. The solution is made upto 50 ml with water and the roots of seedlings in bundles are dipped for 20 min before transplanting.

Sett treatment: Treat the sugarcane setts in 0.05% malathion for 15 minutes to protect them from scales. Treat the sugarcane setts in 0.05% Imidacloprid 70 WS @ 175 g/ha or 7 g/l dipped for 16 minutes to protect them from termites.

Trunk/stem injection: This method is used for the control of coconut pests like black headed caterpillar, mite etc. Drill a downward slanting hole of 1.25 cm diameter to a depth of 5 cm at a light of about 1.5 m above ground level and inject 5 ml of monocrotophos 36 WSC into the stem and plug the hole with cement (or) clay mixed with a fungicide. Pseudo stem injection of banana, an injecting gun or hypodermic syringe is used for the control of banana aphid, vector of bunchy top disease.

Padding: Stem borers of mango, silk cotton and cashew can be controlled by this method. Bark of infested tree (5 x 5 cm) is removed on three sides leaving bottom as a flap. Small quantity of absorbant cotton is placed in the exposed area and 5-10 ml of Monocrotophos 36 WSP is added using ink filler. Close the flap and cover with clay mixed with fungicide.

Swabbing: Coffee white borer is controlled by swabbing the trunk and branches with HCH (BHC) 1 per cent suspension.

Root feeding: Trunk injection in coconut results in wounding of trees and root feeding is an alternate and safe chemical method to control black headed caterpillar, eriophyid mite, red palm weevil. Monocrotophos 10 ml and equal quantity of water are taken in a polythene bag and cut the end (slant cut at 45) of a growing root tip (dull white root) is placed inside the insecticide solution and the bag is tied with root. The insecticide absorbed by root, enter the plant system and control the insect.

Soil drenching: Chemical is diluted with water and the solution is used to drench the soil to control certain subterranean pests. (eg.) BHC 50 WP is mixed with water @ 1 kg in 65 litres of water and drench the soil for the control of cotton/stem weevil and brinjal ash weevil grubs.

Capsule placement: The systemic poison could be applied in capsules to get toxic effect for a long period. (eg.) In banana to control bunchy top vector (aphid) the insecticide is filled in gelatin capsules and placed in the crown region.

Baiting: The toxicant is mixed with a bait material so as to attract the insects towards the toxicant.

Spodoptera: A bait prepared with 0.5 kg molasses, 0.5 kg carbaryl 50 WP and 5 kg of rice bran with required water (3 litres) is made into small pellets and dropped in the field in the evening hours.

Rats: Zinc phophide is mixed at 1:49 ratio with food like popped rice or maize or cholam or coconut pieces (or) warfarin can be mixed at 1:19 ratio with food. Ready to use cake formulation (Bromodiolone) is also available.

Coconut rhinoceros beetle: Castor rotten cake 5 kg is mixed with insecticide.

Fumigation: Fumigants are available in solid and liquid forms. They can be applied in the following way.

Soil: To control the nematode in soil, the liquid fumigants are injected by using injecting gun.

Storage: Liquid fumigants like Ethylene dibromide (EDB), Methyl bromide (MB), carbon tetrachloride etc. and solid fumigant like Aluminium phosphide are recommended in godowns to control stored product pest.

Trunk: Aluminium phosphide 7f to I tablet is inserted into the affected portion of coconut tree and plugged with cement or mud for the control of red palm weevil

V. PLANT PROTECTION APPLIANCES

A. Dusters

Appliances that are used for applying dry dust formulations of pesticides are called as dusters. They make use of an air stream to carry the chemicals in finely divided and dry form to the plants. The dusters consists essentially of a hopper which contains an agitator, an adjustable orifice or metering mechanism and delivery tube. A rotary fan or a below provide the conveying air. They may be operated either manually or by power.

Fig. 28: Farmer operating rotary duster Source: own photograph Chetan jawale ©

1. Rotary dusters: They are also known as crank dusters and fan type dusters. They vary in design and may be shoulder mounted, back mounted or belly mounted. a rotary duster consists of a blower with gear box and a hopper with a capacity to hold 4-5 kg dust. The duster is operated by rotating a crank and the motion is transmitted through the gear to the blower. Generally an agitator is connected to one of the gears. the air current produced by the blower draws the dust from the hopper and discharges out through a delivery tube which may have one or two nozzles (Fig. 28). They are used for dusting field crops, vegetables and small trees and bushes in orchards. The efficiency is 1 to 1.5 per day.

B. Sprayers

Principle: The function of a sprayer is to atomize the spray fluid into small droplets and eject it with some force. The important parts are tank, pump, agitator, pressure gauge, valves, filters, pressure chamber, hose, spray lance, cut of device, boom and nozzle.

1. Tank: To hold the spray fluid during spraying, a sprayer should have a built in or separate container. In case of knapsack and power sprayers the capacity of the tank varies from 9 to 13 litres.

2. Pump: The pump is necessary for creating the energy required for atomization of spray fluid. It is most vital part of a sprayer. A sprayer may be equipped with one of the following types of pumps.

a) Air pump: (Pneumatic): Mostly used in compression sprayers. In this the force created by pump acts, over the spray fluid and the pump does not act directly over the spray fluid.

b) Positive displacement pumps (Plunger, rotary and centrifugal pump): This pump takes a definite volume of liquid inlet and transfer it without any escape to outlet.

3. Agitator: Most of the sprayers are provided with an agitator for dispersing the pesticide uniformly. It may be hydraulic or mechanical agitation.

4. Pressure gauge: It is connected to the pipe line near the nozzle usually.

5. Valves: They govern the direction of the flow of the spray fluid.

6. Filter: Usually this is provided between tank and the pump unit, pump and spray lance and within the lance. This is provided mainly to protect the pump from abrasion, to avoid interference with the function of valves and to prevent blocking of nozzles.

7. Pressure chamber: It is present in sprayers working with hydraulic pumps. It prevents fluctuation in the pressure and effects uniformly in spraying.

8. Hose: It is attached to the sprayer on one end and the spray lance on the other. Mainly plastic and nylon materials are used since they are cheap and light.

9. Spray lance: The nozzle of sprayer is usually attached to a brass rod of variable design. Known as the spray lance the length varies from 35 to 90 cm.

It is usually detachable. In certain cases, it has a 120oC bend to from a goose neck which is useful for spraying under surface of leaf.

10. Cut-off valve: It is used to shut off the liquid. This may be operated by a knob or spring active (trigger cut-off). Three types are used (a) Wheel cut-off valve with strainer. (b) Trigger cut-off valve with strainer. (c) Trigger cut off valve without strainer.

11. Spray boom: Spray bars carrying more than one nozzle is known as spray booms.

12. Nozzle: It breaks the liquid into droplets and spread them into spray droplets.

Fig. 29: Knapsack manual sprayer Source: own photo Chatan Jawale ©

Types of sprayers

Manually operated hydraulic sprayers: In this type, the hydraulic pump directly acts on the spray fluid and discharges it.

Manually operated compression sprayers: These are also known as pneumatic sprayers because air pressure is employed for forcing the liquid through the nozzle for atomization. The containers of these sprayers should not be filled completely with the spray fluid. a part of the container is kept empty so that adequate air pressure can be developed over the spray fluid in the tank.

Pneumatic knapsack sprayer: These sprayers are similar to compression hand sprayers but are adapted for spraying large quantities of liquid (9-10 litres). It comprises of a tank for holding the spray fluid with compressed air, a vertical air pump with a handle, a filler hole, a spray lance with a nozzle and cut off device. Before starting the sprayer, air is compressed into the empty space in the tank. As the spray continues the pressure drops continuously.

a) Body - piece of brass, one end has internal threads and if the threads are inside they will be called as female nozzles and it present outside as male nozzles. One the other end these threads are always on out side.

b) Cap: It is a nut screwed on the body which holds the strainer, orifice plate, washer and swirl plate in position.

c) Swirl plate: Nozzle has a specially drilled swirl plate to give a definite characteristic spray pattern.

d) Washer (sealer): They are of various thickness to allow variation in depth of the swirl chamber and it also prevents the leakage of spray fluid.

e) Stainer: The nozzle is equipped with a strainer. Openings in the strainer are small to prevent the entry to bigger size particle.

HAZARDS CAUSED BY PESTICIDES

The adverse effect caused by pesticides to human beings during manufacture, formulation, application and also consumption of treated products is termed as the hazard.

Pesticide hazard occurs at the time of

a. Manufacturing and formulation

b. Application of pesticides

c. Consumption of treated products

Examples of hazards caused by pesticides

1. In Kerala, in 1953, 108 people died due to parathion poisoning

2. 'Bhopal Gas Tragedy' in 1984 at Bhopal where the gas called Methyl isocyanate (MIC) (an intermediate involved in manufacture of carbaryl) leaked killing 5000 people and disabling 50,000 people. Totally 2,00,000 persons were affected. Long term effects like mutagenic and carcinogenic effects are felt by survivors.

3. Cases of Blindness, Cancer, Liver and Nervous system diseases in cotton growing areas of Maharashtra where pesticides are used in high quantity.

4. Psychological symptoms like anxiety, sleep disturbance, depression, severe head ache in workers involved in spraying DDT, malathion regularly.

5. Endosulfan - causing problem due to aerial spraying in cashew in Kerala - recent controversy - yet to be studied in detail.

Safe handling of pesticides

1. Storage of pesticides

a) Store house should be away from population areas, wells, domestic water storage, tanks.

b) All pesticides should be stored in their original labeled containers in tightly sealed condition.

c) Store away from the reach of children, away from flames and keep them under lock and key.

2. Personal protective equipment

a) Protective clothing that covers arms, legs, nose and head to protect the skin.

b) Gloves and boots to protect hands and feet.

c) Helmets, goggles and facemask to protect hair, eyes and nose.

d) Respirator to avoid breathing dusts, mists and vapor.

3. Safety in application of pesticides

Safe handling of pesticides involves proper selection and careful handling during mixing and application.

a) Pesticide selection

Selection of a pesticide depend on the type of pest, damage, losses caused, cost etc.

b) Safety before application

i. Read the label and leaflet carefully.

ii. Calculate the required quantity of pesticides.

iii. Wear protective clothing and equipment before handling.

iv. Avoid spillage and prepare spray fluid in well ventilated area.

v. Stand in the direction of the wind on back when mixing pesticides.

vi. Do not eat, drink or smoke during mixing.

vii. Dispose off the containers immediately after use.

c) Safety during application

i. Wear protective clothing and equipment.

ii. Spray should be done in windward direction. iii. Apply correct coverage.

iv. Do not blow, suck or apply mouth to any spray nozzle.

v. Check the spray equipment before use for any leakage.

d) Safety after application

i. Empty the spray tank completely after spraying.

ii. Avoid the draining the contaminated solution in ponds, well or on the grass where cattle graze.

iii. Clean the spray equipment immediately after use.

iv. Decontaminate protective clothing and foot wear.

v. Wash the hands thoroughly with soap water, preferably have a bath.

vi. Dispose off the containers by putting into a pit.

vii. Sprayed field must be marked and unauthorized entry should be prevented.

First aid: In cane of suspected poisoning; call on the physician immediately. Before calling on a doctor, first aid treatments can be done by any person.

Swallowed poison

i. During vomiting, head should be faced downwards.

ii. Stomach content should be removed within 4 h of poisoning.

iii. To give a soothing effect, give either egg mixed with water, gelatin, butter, cream, milk, mashed potato.

iv. In case of nicotine poisoning, give coffee or strong tea.

Skin contamination

i. Contaminated clothes should be removed.

ii. Thoroughly wash with soap and water.

Inhaled poison

i. Person should be moved to a ventilated place after loosing the tight cloths.

ii. Avoid applying frequent pressure on the chest.

REFERENCE TABLE

1. Bajwa, W.I. and M. Kogan. "A Collection of IPM Definitions and their Citations in Worldwide IPM Literature." Consortium for International Crop Protection (CICP), New York State Agricultural Exp. Station, Geneva, New York, and Integrated Plant Protection Center (IPPC) Oregon State University, Corvallis, Oregon. 1996. http://www.ippc.orst.edu/cicp/IPM.htm.
2. Benbrook, C.M., and E. Groth III. "Indicators of the Sustainability and Impacts of Pest Management Systems." AAAS 1997 Annual Meeting. Seattle, WA. Feb. 16, 1996.
3. Boutwell, J.L. and R.F. Smith. "A New Concept in Evaluating Pest Management Programs." *Entomological Society Bulletin.* 27(1981):117-18.
4. Dhaliwal, G.S and Ramesh Arora. 1996. Principles of insect pest management. National Agricultural Technology Information Centre, Ludhiana.
5. El-Zik, K.M., and R.E. Frisbie. "Integrated Crop Management Systems for Pest Control." In *CRC Handbook of Pest Management in Agriculture.* Volume III. pp. 3-104. Ed. D. Pimentel, Boca Raton, FL: CRC Press, Inc. 1991.
6. Environmental Protection Agency (EPA). "EPA for Your Information. Prevention, Pesticides and Toxic Substances." (H7506C). 1993, 2 pp.
7. Fernandez-Cornejo, J., S. Jans, and M. Smith. "Pesticide Economic Issues: A Review Article." *Review of Agricultural Economics.* 20(2)(1998):462-88.
8. Glass, E.H. "Pest Management: Principles and Philosophy." In *Integrated Pest Management,* J.L. Apple and R.F. Smith, eds. Plenum Press. New York. 1976, 200 pp.
9. Harper, J.K., M.E. Rister, J.W. Mjelde, B.M. Drees, and M.O. Way. "Factors Influencing the Adoption of Insect Management Technology." *American Journal of Agricultural Economics.* 72(1990): 9971005.
10. Hokkanen, H.M.T. "New Approaches in Biological Control." In *CRC Handbook of Pest Management in Agriculture.* Vol. II. Ed. D. Pimentel and A.A. Hanson. Boca Raton, FL: CRC Press. 1991.
11. National Research Council (NRC). Environmental Studies Board. "Contemporary Pest Control Practices and Prospects: The Report of the Executive Committee." Volume I of *Pest Control:*
12. Jayaraj, J. 1993. Biopesticides and integrated pest management for sustainable crop production. In: Agrochemicals in sustainable agriculture (Ed. N.K. Roy). New Delhi, APC Publications.

13. Khader Khan, H. 1996. Integrated pest management and sustainable agriculture. Farmers and Parliament. 30 (2):15-17.
14. Birthal, P. S. 2003. Economic Potential of Biological Substitutes for Agrochemicals
15. Office of Technology Assessment (OTA). *Pest Management Strategies,* Vols I & II. Congress of the United States. Washington, DC. 1979.
16. Singh, V. and M.P. Ghewande. 1980. Plant protection in oilseeds. Pesticides
17. Annual 1980-81, 30-35.
18. *An Assessment of Present and Alternative Technologies.* (Five Volumes). Washington, DC: National Academy of Sciences. 1975, 506 pp.
19. Pedigo, L.P. "Economic Thresholds and Economic Injury Levels." In E. B. Radcliffe and W. D. Hutchison (eds.), Insect Pest Management: Radcliffe's IPM World Textbook, URL: http://ipm-world.umn.edu. University of Minnesota, St. Paul, MN.
20. Sailer, R.I. "Extent of Biological and Cultural Control of Insect Pests of Crops." In *CRC Handbook of Pest Management in Agriculture.* Volume II. pp. 1-12. Ed. D. Pimentel, Boca Raton, FL: CRC Press, Inc. 1991a.
21. Subramaniam, K.V. 1997. Critical factors in adoption of integrated pest management technologies: An economic analysis. In: First National Symposium in Pest Management in Horticulture Crops. IIHR, Bangalore. Oct 15-17.
22. U.S. Department of Agriculture, Economic Research Service. *Extent of Spraying and Dusting on Farms, 1958 with Comparisons.* SB-314. May 1962.
23. Fred Baur. Insect Management for Food Storage and Processing. American Association of Cereal Chemists. ISBN 0-913250-38-4.
24. Alan M. Bowerman & Joe E. Brooks (1971). "Evaluation of U-5897 as a male chemosterilant for rat control". Journal of Wildlife Management **35** (4): 618–624. doi:10.2307/3799765. JSTOR 3799765.
25. Wright, MG; Hoffmann, MP; Kuhar, TP; Gardner, J; Pitcher, SA (2005). "Evaluating risks of biological control introductions: A probabilistic risk-assessment approach". Biological Control **35**: 338–347. doi:10.1016/j.biocontrol.2005.02.002.
26. Wajnberg, E., Hassan, S.A. (1994). Biological Control with Egg Parasitoids. UK: CABI Publishing.
27. Kaya, Harry K. et al. (1993). "An Overview of Insect-Parasitic and Entomopathogenic Nematodes". In Bedding, R.A. Nematodes and the Biological Control of Insect Pests. CSIRO Publishing. ISBN 978-0-643-10591-1.

28. Flint, Maria Louise & Dreistadt, Steve H. (1998). Clark, Jack K., ed. Natural Enemies Handbook: The Illustrated Guide to Biological Pest Control. University of California Press. ISBN 978-0-520-21801-7.

Online Reference

[ZOOLOGY DEPARTMENT H.P.T.Arts & R.Y.K Science College ...](#)

zoologyryk.blogspot.com/p/agriculture-entomology.html

The author of this blog is Dr. Chetan Jawale.